SpringerBriefs in Applied Sciences and Technology

PoliMI SpringerBriefs

W0235299

More information about this series at http://www.springer.com/series/11159
http://www.polimi.it

Matteo Vincenzo Rocco

Primary Exergy Cost of Goods and Services

An Input–Output Approach

**POLITECNICO
DI MILANO**

Matteo Vincenzo Rocco
Department of Energy
Politecnico di Milano
Milan
Italy

ISSN 2191-530X   ISSN 2191-5318   (electronic)
SpringerBriefs in Applied Sciences and Technology
ISSN 2282-2577   ISSN 2282-2585   (electronic)
PoliMI SpringerBriefs
ISBN 978-3-319-43655-5   ISBN 978-3-319-43656-2   (eBook)
DOI 10.1007/978-3-319-43656-2

Library of Congress Control Number: 2016947758

Printed on acid-free paper

This Springer imprint is published by Springer Nature
The registered company is Springer International Publishing AG Switzerland

Preface

Among the large multiplicity of natural resources, non-renewable primary fossil fuels, namely coal, crude oil and natural gas, play a crucial role in sustaining the human economies and their production activities. Nowadays, scarcity of stocks and emissions of pollutant and greenhouse gases are the main concerns related to fossil fuels depletion, and many political and research initiatives are facing these issues by promoting a rational use of primary energy. This objective could be achieved acting in two main ways: (1) fostering a transition toward alternative and more efficient energy conversion systems (e.g., increase efficiency of conventional power systems, or increase penetration or renewables); (2) reducing the primary energy *embodied in goods and services produced by national economies.*

Primary energy *embodied* in goods and services, also called *energy cost,* is defined as the amount of primary fossil fuels *directly* and *indirectly* required to deliver such products. Reduction of the embodied energy of products can be performed by developing innovative production policies or by improving production processes and supply chains. However, in order to check whether these interventions result in a net reduction of the overall primary energy requirements, an accurate evaluation of the energy embodied in products is crucial. The domain of the second activity lies in the field of *Life Cycle Assessment* (LCA), and encompasses both technological and economic aspects related to supply chains operation.

This book introduces and formalizes the *Exergy-based Input–Output framework* (ExIO), a comprehensive methodology useful to account for the primary fossil fuels required for the production of goods and services within a given national economy. The main features of the developed framework are aimed at dealing with the flaws of current state-of-the-art methodologies: (1) *Exergy* is here assumed as the unique metric for fossil fuels characterization and for energy conversion systems analysis; (2) *Input–Output analysis* is assumed as the computational structure of ExIO: this approach relies on standard and freely available data sources, making its application simpler and faster with respect to other process-based methods; (3) the *Hybrid* approach have been developed to increase the accuracy of results and to analyze energy conversion systems, assessing and maximizing their overall thermodynamic efficiencies based on suited indicators; (4) the *Bioeconomic ExIO model* has been finally proposed to internalize the effects of human labour within the primary embodied energy of products.

The book is the result of 3 years of research carried out at the Department of Energy, Politecnico di Milano: it ought to be a step forward in the fields of *Thermodynamics methodologies for system analysis* and *Life Cycle Assessment*.

I am grateful to my Ph.D. Supervisors Prof. Emanuela Colombo and Prof. Fabio Inzoli. I am especially grateful to my wife Chiara, my parents, and all the friends and colleagues for their support during the development of this book.

Milan, Italy   Matteo Vincenzo Rocco
June 2016

Contents

Symbols, Acronyms, and Abbreviations

Note about the Adopted Notations

a, A, a, A	Scalars
$\boldsymbol{a}, \boldsymbol{A}, \mathbf{a}, \mathbf{A}$	Vectors, matrices
$\mathbf{i}, \mathbf{i}\,(n \times m)$	Vector of 1s, with n rows and m columns
$\mathbf{I}, \mathbf{I}_n$	Identity matrix of order n
$\mathbf{0}\,(n \times m)$	$n \times m$ empty matrix
$\text{diag}(\mathbf{a})\hat{\mathbf{a}}$	Diagonal matrix, with elements of the $\mathbf{a}$ Vector as diagonal elements
$\mathbf{a}^{\mathrm{T}}, \mathbf{A}^{\mathrm{T}}$	Transposed vector and matrix
$\mathbf{A}^{-1}$	Inverse matrix
$a, \bar{a}$	Per mass unit, per mole unit
$\text{row}_i\,(\mathbf{A})$	Identify the ith row of matrix $\mathbf{A}$
$\text{column}_i\,(\mathbf{A})$	Identify the ith column of matrix $\mathbf{A}$

Symbols

$\text{ex}_i, \mathbf{ex}$	Exports (for ith sector and total) (€)
$\text{f}_i, \mathbf{f}$	Final demand of the ith process, final demand vector
$\text{g}_i, \text{G}, \text{g},$	Government purchases (for ith sector and total) (€), gravity acceleration (9.81 m/s^2)
h_i, H	Household purchases (for ith sector and total) (€)
i_i, I	Purchases for private investment purposes (for ith sector and total) (€)
l_i, L	Payments for labor compensation (for ith sector and total) (€)
m_i, M	Imported products (for ith sector and total) (€)
n_i, N	Government expenses and other minor voices (for ith sector and total) (€)
$\text{p}_i, \mathbf{p}$	Price of ith product, price vector
$\text{r}_{ki}, \mathbf{r}, \mathbf{R}$	Exogenous transactions of the kth resources to the ith sector, exogenous transactions vector and matrix

x_i, $\mathbf{x}$	Total endogenous transactions of ith process, total endogenous transactions vector
x_{ii}, $\mathbf{Z}$	Endogenous transactions from ith to jth, endogenous transactions matrix
$\mathbf{C_{NS}}$, $\mathbf{C_{SN}}$	Upstream cutoff matrix, downstream cutoff matrix
E_i, $\mathbf{E}$	Total embodied exogenous transaction coefficient, total embodied exogenous transaction matrix
$\dot{Q}$, Q	Heat rate (W), heat (J)
a_i, $\mathbf{A}$	Technical coefficient, technical coefficients matrix
b_i, $\mathbf{B}$	Intervention coefficient, intervention coefficients matrix
c_p, c_v	Isobaric heat capacity (J/kgK), isochoric heat capacity (J/kgK)
e_i, $\mathbf{e}$	Specific embodied exogenous transaction coefficient, specific embodied exogenous transaction matrix
l_i, $\mathbf{L}$	Leontief inverse coefficient, Leontief inverse coefficients matrix
$\dot{m}$, $\dot{n}$	Mass flow rate (kg/s), molar flow rate (mol/s)
η_{ex}	Exergy functional efficiency
h, $\mathbf{h_W}$	Hours (h), working hours requirements vector,
h, $\dot{H}$, H	Specific enthalpy (J/kg), enthalpy rate (W), enthalpy (J)
$\mathbf{I}$	Identity matrix
$\mathbf{i}$	Summation vector
M2	Monetary circulation (€)
n	Number of productive processes (–)
p	Pressure (Pa)
R	Universal gas constant (8,314 J/kmolK)
t, T	Time (s), temperature (K)
V	Volume (m^3)
v, $\mathbf{v}$	Values added (€), value-added vector
w, $\dot{W}$, W	Velocity (m/s), work rates (W), work (J)
X	Total outlays of the nation (€)
y	Mass fraction (g/g)
z	Elevation (m)
HV	Heating Value (J/kg)
e, $\dot{E}$, E	Specific energy (J/kg), power (W), energy (J)
ee, $\dot{EE}$, EE	Specific extended exergy (J/...), extended exergy rate (W), extended exergy (J)
ex, $\dot{Ex}$, Ex	Specific exergy (J/kg), exergy rate (W), exergy (J)
g, $\dot{G}$, G	Specific Gibbs function (J/kg), Gibbs function (W), Gibbs function (J)
s, $\dot{S}$, S	Specific entropy (J/kgK), entropy rate (W/K), entropy (J/K)
x	Mole fraction (mol/mol)
β	Szargut factor (–)
$\lambda(\mathbf{A})$	Eigenvalues of matrix $\mathbf{A}$

μ	Chemical potential (J/kmolK)
ν	Stoichiometric coefficient (–)
ρ	Spectral radius (–)

Subscripts

0	Environmental state
00	Dead State
B	Bioeconomic
C	Closed
ch	Chemical
D	Destruction
D	Destruction
D	Direct
env	Environment
ex	Exergy
ext	Externalities
f	Final Demand
gen	Generation
H	Hybrid, Households
I	Indirect
K	Capitals
kn	Kinetic
L	Labor
L	Losses
LC	Life Cycle
mix	Mixture
N	National
O	Environmental Remediation
P	Products
ph	Physical
pt	Potential
Q	Heat Interaction
R	Reactants
rev	Reversible
S	System
tot	Total
W	World
Wh	Working Hours
Z	Intermediate Transactions

Chapter 1
Introduction

Traditionally, the increase of energy efficiency of productive systems was driven by the search for the attainment of the maximum useful product with the minimum consumption of resources.

Until recently, the evaluation of the energy-resources consumption of productive systems encompasses the energy flows that are *directly* absorbed by the system under consideration during its operating life. Today, the concept of energy-resources consumption is undergoing a radical re-evaluation, in response to the acknowledged interdependency of the productive sectors with both the environment and the society. Indeed, modern economies are sustained by flows of fossil fuels extracted from natural environment: it can be said that *all* the goods and services produced within a given economy are characterized by an *embodied energy* (also called *cumulative energy*, *primary energy* or *energy cost*), which is defined in this book as the *direct* and *indirect* amount of primary fossil fuels required to deliver the considered products.

This chapter aims at clarifying the concept of *natural resources*, highlighting the relevance that non-renewable energy-resources have in the economic process. The emerging needs of resource accounting methods are finally identified and the objectives of the work are defined.

1.1 Role of Natural Resources in the economic process

Natural resources can be defined as all the goods provided by nature that are useful in order to sustain human economic systems. Economic production consists the transformation of such natural resources into something of values for humans, that is, something that creates welfare, quality of life, utility or whatever else provide us satisfaction. All the productive processes, even the production of immaterial services, require natural resources to sustain their production, and inevitably generate wastes.

© The Author(s) 2016
M.V. Rocco, *Primary Exergy Cost of Goods and Services*,
PoliMI SpringerBriefs, DOI 10.1007/978-3-319-43656-2_1

In last decades, many scientists argued that the nature of economic and biological processes are similar: the human economy can be productive and satisfy human needs *only* by transforming available raw materials and energy into unavailable flows of wastes. However, Traditional economic paradigms have not paid much attention to the physical roots of economic production, according to the assumption that the biophysical world does not constrain the development of the economic systems (Daly and Farley 2010).

The issues of natural resources scarcity and the environmental effects due to their exploitation lead public opinion and policymakers in recognize that the growth of modern economies is physically constrained (Cleveland and Ruth 1997; Costanza 1992; Costanza and Daly 1987; Mayumi 2009). In last decades, attention has been devoted to the thermodynamic limits of the human economy, and the *entropic* nature of the economy has been recognized, as first emphasized by *Nicolas Georgescu-Roegen* is his *The Entropy law and the Economic Process* (1971) (Bakshi et al. 2011; Mayumi 2009). Nowadays, efficiency in resources use become one of the main goal of political initiatives such as the *European 2020 strategy* (EU 2011).

1.1.1 Primary and secondary factors of production

Since natural resources are directly harvested from the natural environment, literature usually refer to them as the *primary factors of production* or *natural capital*. The following general classification can be found in literature (Cleveland and Ruth 1997; Daly and Farley 2010):

- *Fossil fuels*. Raw coal, crude oil, natural gas and nuclear fuels are part of this category. All of these fuels are classified as non-renewable resources, since the rate of their extraction by the world economies results faster than the rate at which they are reproduced by natural processes. Our modern societies are strongly dependent on this fixed stock of energy, and the quantification of its total amount is extremely difficult;
- *Mineral resources*. They are represented by the highly concentrated stocks of metals and minerals ores in the Earth crust. They are classified as non-renewable resources, because even if they can be recycled, it has been demonstrated that their total recycling is theoretically impossible (Ayres 1999);
- *Water*. In contrast to fossils, minerals and metals stocks, water resources are renewable as a result of the hydrologic cycle;
- *Land*. It is defined as the soil that supports physical human infrastructures and soil that can be used for agriculture;
- *Solar energy*. This category encompasses solar radiation as well as its derivatives, such as kinetic energy of the wind, potential energy of water and biomass. Obviously, solar energy is defined as renewable.

Based on their physical and chemical qualities, material resources *consumption* can be classified in two sub-categories: resource *use* or *depletion*. A resource is said to be *used* if it is possible to reuse it again after the consumption process: minerals, water and land are often part of this category. On the other hand, if the consumption of a resource imply a radical change of its chemical and physical properties, as for the combustion process of fossil fuels, resources are said to be *depleted* (Bakshi et al. 2011).

Natural resources sustain economic production, allowing to produce *energy carriers*, *goods*, *services*, *monetary capitals* and *labour* (*working hours*): since these products are essential for economic production, they can be simultaneously considered as input and output of the human economy, as depicted in Fig. 1.1. For such reason, these contributions are here defined as the *secondary factors of production*. The distinction among primary and secondary factors of production is crucial for the development of primary resource accounting methods and it will be very useful in further chapters.

1.1.2   Focus on non-renewable Energy-Resources

The crucial role of energy in modern economic activities is undeniable (Giampietro et al. 2012; Wilting 1996). Already in 1933, *Soddy* wrote: «*If we have available energy, we may maintain life and produce every material requisite necessary. That is why the flow of energy should be the primary concern of economics.*» (Costanza 1980; Wilting 1996). Based on IEA data, nowadays fossil fuels provide more than 75 % of the total world energy consumption: as stated by *Odum*, the prosperity and stability of modern societies are then inextricably linked to the production and consumption of fossils, namely *raw coal, crude oil* and *natural gas* (Hall et al. 2014; Odum 1973).

Production activities and requirements of energy are strictly related, as shown in Fig. 1.2, where the trends of the World's *Gross Domestic Product* (GDP) and *Total Primary Energy Supply* (TPES) per capita in the period 1971–2011 are compared.

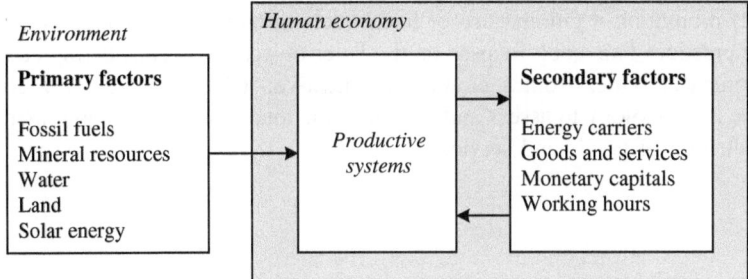

Fig. 1.1 Primary and secondary factors of production

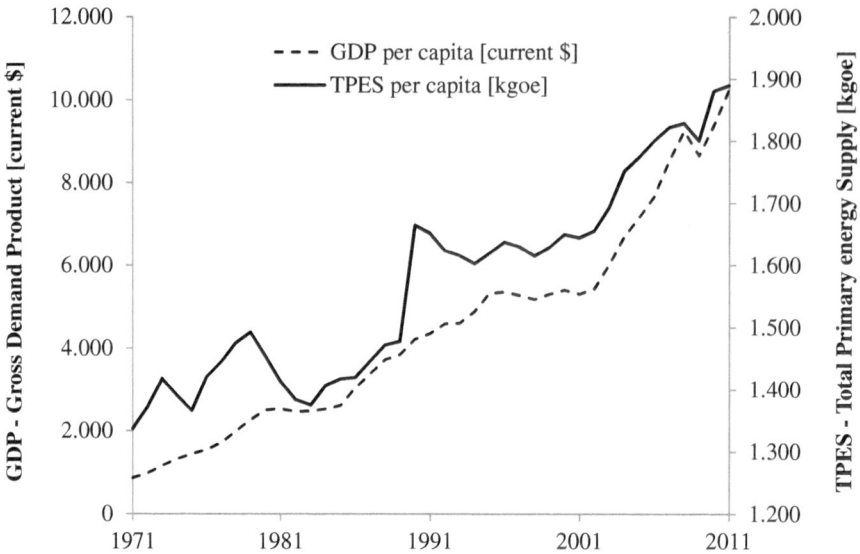

Fig. 1.2 Trend of Gross Domestic Production per capita (GDP) and Total Primary Energy Supply per capita (TPES) for the World in the period 1971–2011 (IEA data)

Two main issues are connected with depletion of fossil fuels:

- *Scarcity of stocks.* Since stocks of fossil fuels are finite, the depletion of the presently known global resources which are economically available will probably become reality sometime in the current century (Bardi 2011; Costanza 1991; Daly and Farley 2010; Mayumi 2009);
- *Pollutants and greenhouse gases emissions.* Extraction, refinements and combustion of fossil fuels cause emissions of greenhouse gases (mainly CO_2) and pollutants. Emissions due to hydrocarbons are dangerous for the health of humans and ecosystems, and they represent a contribution in raising up the concentration of greenhouse gases in the atmosphere (Solomon 2007).

Political initiatives as well as research efforts are thus facing these issues acting in two main directions: (1) claiming a *transition* towards alternative power sources and (2) promoting a rational use of fossil fuels through the so-called *energy efficiency* practice. This book focuses on the latter objective by considering economic and human activities from a *physical* perspective: it aims to provide a comprehensive methodology to assess and to reduce the total fossil fuels requirements due to production of goods and services.

1.2   Emerging needs in Environmental Impact Analysis

The ultimate purpose of the economy is to satisfy the households' demand of goods and services. Modern economies can be represented as intricate networks of productive processes connected to each other by flows of material goods, energy and nonmaterial services. Such networks are connected to the natural environment, taking natural resources and rejecting wastes. The acknowledged interdependency of the productive sectors with both the environment and the society at large makes quantitative evaluation of environmental impact due to goods and services production a very challenging task. Indeed, the production of all the goods and services are sustained, in a both direct and indirect way, by flows of resources extracted from natural environment.

Environmentally conscious political and technological decisions require to know their total effects in terms of primary fossil fuels requirements: literature clearly states that without a proper evaluation of the *overall* resource consumption of one specific productive system, capable to include also the indirect supply chains requirements, misleading results may be obtained. For this reason, many efforts are focused on the definition of methods to account for the overall fossil fuels contribution to individual products (Ayres 2008; Ayres et al. 2007, 2013; Liao et al. 2012; Suh 2009).

From the analyst perspective, the following four gaps emerge from the literature.

1. *Identification of a standardized accounting method*: consensus about the most appropriate resources accounting scheme is still nonexistent. Indeed, Life Cycle based methods are not defined in an unique and unambiguous way, and they rely on extensive data collection procedures, resulting in large uncertainties in results;
2. *Energy-resources characterization*: identification of one comprehensive thermodynamic-based metric for energy-resources characterization is claimed;
3. *Evaluation of efficiency of energy systems in a Life Cycle perspective*: performances evaluation of energy conversion systems by means of traditional First and Second Law indicators neglect the indirect effects linked to the consumption of non-energy related products and externalities. Novel thermodynamic based indicators are claimed to obtain useful insight for the optimization procedure;
4. *Role of externalities*: the role that externalities of productive systems (and human labour among others) have in primary resources consumption needs to be quantitatively clarified.

1.3   Objectives and structure of the book

The general objective of this book is the *development of a resource accounting method for the evaluation of non-renewable primary energy-resources requirements of any product of a given economy*. This has been reached through the definition and the formalization of the *Exergy based Input-Output analysis* (ExIO), a comprehensive framework that integrates *Input-Output analysis* and *Exergy analysis* for the thermodynamic performance evaluation and optimization of energy systems in a Life Cycle perspective.

Specific objectives of the book stem from the emerging issues previously highlighted:

1. The method is formalized in order to allow *reproducible, reliable* and *accurate* resource accountings in a *standard* and *simple* way. In this work, the mathematical formulation of ExIO is based on *Input-Output analysis*, which allows to define standardized time and space boundaries for any analyzed system or product, encompassing its whole Life Cycle. The approach relies on standard and freely available data sources, avoiding extensive data mining processes and making the application of the analysis simpler and faster with respect to traditional *process based* LCA;

2. In this work, *Exergy* is assumed by ExIO as the suited thermodynamic-based metric for fossil fuels characterization and for energy conversion systems analysis;

3. The ExIO method should be able to analyze energy conversion systems in detail, proposing a set of indicators useful for optimization purposes. In this work, accuracy of results can be selectively increased through the *Hybrid-ExIO analysis*, which allows to focus on detailed systems operating inside broader sectors of the economy;

4. The Bioeconomic ExIO model has been proposed to internalize the effects of *human labour* consumption within the primary energy-resources accounting of goods and services;

After the brief overview about the concept of natural resources and their role in the economic process provided by this chapter, the rest of the book is arranged as follows.

Chapter 2 presents the state of the art of resources accounting techniques: mathematical structure of *Input–Output analysis* (IOA) and *Process Analysis* (PA) are formalized for a generic productive system, compared and finally discussed. What emerges from this chapter is that any resources accounting problem may be described in a more efficient, simple and standardized way through IOA rather than using PA.

In Chap. 3, a critical literature review about the *Thermodynamic based methods for system analysis* is provided. *Energy, Entropy* and *Exergy* based Life Cycle methods are comparatively analyzed and a taxonomy is proposed. Finally, the use of such metrics for energy-resources characterization is discussed. What emerges from this chapter is that exergy is widely considered as the most suited numeraire to account for energy-resources consumption, but the application of Exergy to Life Cycle Assessment methods requires further methodological improvements.

Chapter 4 is the core of this book: it merges the resources accounting technique of IOA with the concept of exergy, formalizing the *Exergy based Input Output analysis* (ExIO). This method allows to evaluate the primary exergy cost of goods and services produced by a specific national economy. The method relies on *Monetary Input–Output Tables* (MIOTs) of national economies as standardized and constantly updated data source. The main methodological achievements of this chapter can be summarized as follows: (1) a complete formalization of ExIO method is proposed. Specifically, different techniques are proposed to account for international trades of products; (2) The Hybrid ExIO approach is proposed and formalized in order to increase the accuracy of results obtained through the use of standard ExIO analysis and to perform Life Cycle Assessment of detailed products and systems; (3) The Hybrid ExIO approach is adapted in order to perform Thermoeconomic analysis and Design Evaluation of energy conversion systems.

In Chap. 5 the role of human labour in primary energy-resources accounting is investigated, and the *Bioeconomic ExIO model* is proposed and formalized to account for the effects that working hours consumption has on primary energy-resources invoked for the production of goods and services.

Finally, in Chap. 6 the ExIO framework is applied to the following case studies: (1) evaluation of primary exergy cost of goods and services produced by different *national economies*; (2) application of Hybrid ExIO model for Thermoeconomic analysis and Design Evaluation of an Italian *Waste to Energy power plant* in a Life Cycle perspective; (3) application of standard and Bioeconomic ExIO for the detailed analysis of the Italian economy, and for the comparative analysis of the primary exergy costs of dishwashing alternatives.

Conclusions of the book remark its main achievements and also gives a perspective about the future possible research paths.

References

Ayres, R. U. (1999). The second law, the fourth law, recycling and limits to growth. *Ecological Economics, 29*, 473–483.

Ayres, R. U. (2008). Sustainability economics: Where do we stand? *Ecological Economics, 67*, 281–310.

Ayres, R. U., Turton, H., & Casten, T. (2007). Energy efficiency, sustainability and economic growth. *Energy, 32*, 634–648.

Ayres, R. U., van den Bergh, J. C. J. M., Lindenberger, D., & Warr, B. (2013). The underestimated contribution of energy to economic growth. *Structural Change and Economic Dynamics, 27*, 79–88.

Bakshi, B. R., Gutowski, T. G. P., Gutowski, T. G. P., Sekulic, D. P., & Sekulić, D. P. (2011). *Thermodynamics and the destruction of resources.*

Bardi, U. (2011). *The limits to growth revisited.* Springer.

Cleveland, C. J., & Ruth, M. (1997). When, where, and by how much do biophysical limits constrain the economic process?: A survey of Nicholas Georgescu-Roegen's contribution to ecological economics. *Ecological Economics, 22*, 203–223.

Costanza, R. (1980). Embodied energy and economic valuation. *Science, 210*, 1219–1224.

Costanza, R. (1991). Ecological economics: A research agenda. *Structural Change and Economic Dynamics, 2*, 335–357.

Costanza, R. (1992). *Ecological economics: The science and management of sustainability.*

Costanza, R., & Daly, H. E. (1987). Toward an ecological economics. *Ecological Modelling, 38*, 1–7.

Daly, H. E., Farley, J. (2010). *Ecological economics: Principles and applications* (2nd ed.).

EU. (2011). A resource-efficient Europe–Flagship initiative under the Europe 2020 Strategy. *COM (2011)*, 2.

Giampietro, M., Martin, J. R., & Ulgiati, S. (2012). Can we break the addiction to fossil energy? *Energy, 37*, 2–4.

Hall, C. A. S., Lambert, J. G., & Balogh, S. B. (2014). EROI of different fuels and the implications for society. *Energy policy, 64*, 141–152.

Liao, W., Heijungs, R., & Huppes, G. (2012). Thermodynamic analysis of human–environment systems: A review focused on industrial ecology. *Ecological Modelling, 228*, 76–88.

Mayumi, K. (2009). Nicholas Georgescu-Roegen: His bioeconomics approach to development and change. *Development and Change, 40*, 1235–1254.

Odum, H. T. (1973). Energy, ecology, and economics. *Ambio*, 220–227.

Solomon, S. (2007). *Climate change 2007-the physical science basis: Working group I contribution to the fourth assessment report of the IPCC.* 4.

Suh, S., (2009). *Handbook of input-output analysis economics in industrial ecology.*

Wilting, H. C., (1996). *An energy perspective on economic activities.*

Chapter 2
Review of Resources Accounting Methods

Development of practical approaches for the evaluation of environmental impact caused by the production of goods and services is one of the most relevant and debated topic in *Industrial Ecology* (IE). According to the literature, one of the main goal of IE resides in the definition of a comprehensive and unified method to account for resources and wastes embodied in goods and services (Bullard et al. 1978; Hendrickson et al. 1997; Williams 2004). This chapter introduces two mathematical accounting schemes widely adopted in literature: *Input-Output analysis* and *Process analysis*. In this chapter, the analytical structure of both the approaches is formalized.

2.1 Life Cycle Assessment

Consumption of natural resources is closely connected with economic and environmental burdens by the intricate nets of processes by which the modern economy transforms, uses and disposes the inputs and outputs of its production. Therefore, understanding the structure of the economy that governs flows of primary and secondary factors of production between producing industries and consuming households is indispensable for solving the problems of both limited resources availability and emissions of pollutants and greenhouse gases (Suh and Kagawa 2005). In last decades, the discipline of *Life Cycle Assessment* (LCA) emerged to face these issues.

The concept of LCA originated in early 1970s, when the issues related to energy efficiency and consumption of scarce raw materials became relevant, in order to provide a unified framework for the evaluation of the total environmental burdens caused by human activities. In 1990s, the *Society of Environmental Toxicology and Chemistry* (SETAC) defined for the first time formal guidelines for environmental assessment of products. These guidelines were formulated in order to provide a complete and clear picture about the interaction between the analyzed productive

© The Author(s) 2016
M.V. Rocco, *Primary Exergy Cost of Goods and Services*,
PoliMI SpringerBriefs, DOI 10.1007/978-3-319-43656-2_2

system and the natural environment, understanding the overall and interdependent nature of the environmental consequences of human activities and thus providing useful information to decision-makers. After SETAC, attempts to define the LCA framework have been made by the *International Organization for Standardization* (ISO), which started to handle the standardization of the methodology, publishing the well-known standards in the 14040 series (UNI 2001). These days, software tools and databases are specifically developed to perform LCA analysis, and the framework is continuously developing from both theoretical and computational viewpoint (Heijungs and Suh 2002b; Pennington et al. 2004; Peters 2007; Rebitzer et al. 2004; Suh et al. 2004).

The aim of LCA is the evaluation of the direct and indirect environmental burdens connected with the production, use and disposal of a given system (Guinée 2002a; In and Curran 1994; Klöpffer 1997). Environmental burdens associated to a certain product can be distinguished among *loadings* and *impacts*. Loadings are material and energy flows that cross the boundaries of the considered system that are quantitatively measureable, whereas impacts concern the consequences of loadings on environment or human health, and are sometimes considered qualitatively (Guinée 2002b). The assessment includes the entire life cycle of the product or activity, encompassing extraction and processing of raw materials, manufacturing, distribution, use, re-use, maintenance, recycling, final disposal and all the other involved services and treatments. Standard regulation ISO 14040 defines four steps for the application of LCA (UNI 2001):

1. *Goal and Scope definition.* In the first step, objective of the analysis, functional unit, temporal and spatial extension of the system (system boundaries) are defined. Moreover, assumptions, strategies and procedures for data collection are established;
2. *Inventory analysis.* It aims at quantifying inputs and outputs that cross the previously defined boundaries: energy, raw materials, products, co-products and wastes that participate to the life cycle of the functional unit are considered and collected in this phase;
3. *Impact assessment.* In this phase, results of the inventory analysis are translated into potential environmental burdens, mainly related to resources use, human health impacts and ecological impacts;
4. *Results interpretation.* In this last phase, analysts are called to examine results in order to identify different options that can be undertaken in order to reduce environmental burdens. Therefore, this phase could requires to iteratively repeat one or more previous phases.

Fundamental literature about LCA reveals that all the above listed application steps require to be further developed and improved: an exhaustive survey about basics, developments and unsolved problems in LCA can be found in literature (Ekvall and Finnveden 2001; Pennington et al. 2004; Reap et al. 2008; Rebitzer et al. 2004).

Within the theoretical framework of LCA, the issue of primary resources accounting started to be addressed from a methodological standpoint in response to the energy price rising, growing awareness of materials scarcity and negative impact of economic production on the environment (Meadows et al. 1972). The new awareness of the negative aftermath caused by the intensive use of fossil fuels moved the focus of the system analysts from the evaluation and reduction of direct energy requirements to a wider perspective. Several studies about the calculation of direct and indirect energy requirements of products were implemented during these years, and different methods were proposed. For instance, *Chapman* estimated the primary energy consumption (also named *energy cost*) of copper, aluminum and refined oil fuels (Chapman 1974; Chapman et al. 1974; Chapman and Faculty 1975). Over the same period, the first studies about the primary energy require-ments of national production were introduced by *Bullard* in (Bullard et al. 1978), referring to *Hereenden* and *Tanaka* studies on the energy cost of households pur-chases in the U.S. economy (Herendeen and Tanaka 1976). *Bullard* and *Hereenden* quantifying the energy cost of goods and services for energy saving purpose, identifying products that required higher total energy use and proposing their substitution (Bullard and Herendeen 1975). Almost simultaneously, *Wright* esti-mated the primary energy cost of British national production with a similar approach (Wright 1975). A few years later, *Costanza* and *Hereenden* tried to find interrelationships between energy cost and economic cost of goods and services trying to find evidences of the so-called *Energy Theory of Value* (Costanza 1980, 2004; Pokrovskii, 2011).

In last decades, methods for primary resources cost accounting were subjected to a refinement by *Treloar, Suh, Lenzen, Duchin, Hendrickson, Szargut* and other scientists (Dixit et al. 2010; Duchin 1992, 1998; Duchin et al. 1996; Hendrickson et al. 2010; Suh 2009; Suh and Huppes 2005; Suh et al. 2004; Suh and Nakamura 2007; Szargut 2005; Szargut et al. 1988, 2002; Treloar 1997, 1998), and such accounting methods are nowadays applied in various field of economic production: buildings (Fay et al. 2000; Hammond and Jones 2008; Sartori and Hestnes 2007; Treloar et al. 2001), industrial products (Lenzen and Dey 2000), energy systems (Lenzen and Munksgaard 2002), automotive systems (Lave et al. 2000), services (Spielmann and Scholz 2005), and so on.

The accounting methods described in the following represent a unified and comprehensive reformulation of the practical approaches for the application of LCA methodology. Different formalization of the same methods can be found in litera-ture (Heijungs and Suh 2002b; Hendrickson et al. 2010; Joshi 1999; Nakamura et al. 2007; Suh and Huppes 2005).

2.2   Introduction to resources accounting

Production of goods and services by economic and human activities is ultimately sustained by a flow of resources that may be taken from the environment (fossil fuels, solar radiation, water, soil, etc.) or that are produced by other processes (semi-finished products, electricity, etc.). As an instance, the electricity production of a solar photovoltaic system requires solar energy during its operation, but also requires energy due to the production of the solar cells. Therefore, the evaluation of resources *directly* and *indirectly* required (i.e. *embodied*) by a certain system strongly depends on the objectives and the choices of the analyst, which must clearly and unambiguously define the *kind of resources* to be accounted for, as well as the extensions of *time* and *space boundaries* of the analyzed system (Suh 2005; Suh and Huppes 2005). As an instance, one may be interested in quantifying the amount of primary fossil fuels embodied in the energy produced by a fuel cell along its whole life cycle, the total electricity embodied in products of a chain of processes in one year, the working hours embodied in a defined production process in one day, and so on.

Theoretical description of embodied resources accounting methods requires the introduction of some fundamental definitions. Let us consider the simple network of *productive processes* of Fig. 2.1, each of which produces and exchanges the product x in a reference *time window*, say one hour.

The definition of the useful products of the network, as well as the required resources, depends on the choices and needs of the analyst. Resources to be accounted for could be *primary* (i.e. directly taken from the natural environment, such as primary energy-resources, ores, etc.) or *secondary* (i.e. produced by other processes, such as electricity, semi-finished materials, labour, etc.). Let us consider the two following simple cases:

1. *Evaluation of the amount of kth products embodied in the ith products in the given time window.* In this case, the analyst defines the *system* as composed by the *i*th and *j*th processes (shaded area in Fig. 2.2, left side). The *i*th process deliver the product f_i outside the network, defined as the *final demand* (also called *target product* or *functional unit*) and considered as the reason why such network exists. The amount of products exchanged between *j*th and *i*th processes is defined as the *intermediate transaction* x_{ij}, and each process thus absorbs a number of (secondary) *exogenous resources* r_k from the *k*th process defined outside system boundaries. Therefore, the direct and indirect (i.e. the

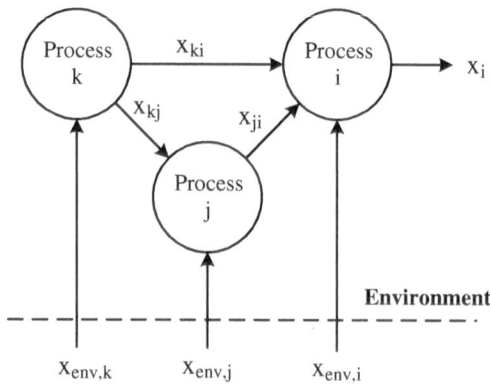

Fig. 2.1 Outline of a generic network of productive processes

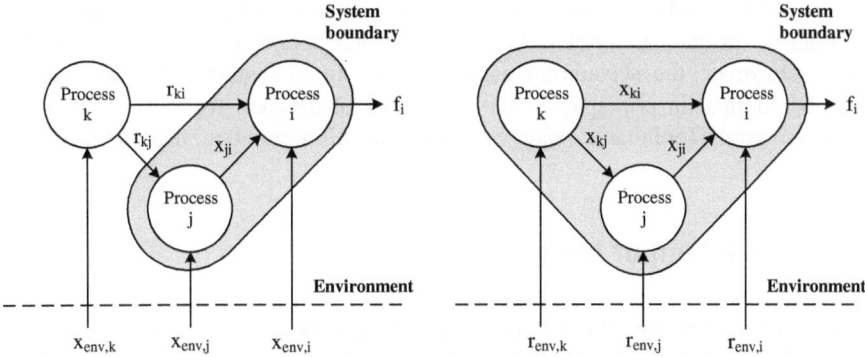

Fig. 2.2 Outline of a generic productive system. The system produces the final demand by absorbing resources from other processes (*left side*) or from the environment (*right side*)

embodied) amount of *k*th resources required to deliver the unit of final demand can be calculated through relation (2.1).

$$e_i = \left(r_{ki} + r_{kj}\right)/f_i \tag{2.1}$$

2. *Evaluation of the amount of primary resources embodied in the ith products in the given time window.* In this case, the analyst defines the *system* as composed by all the processes that absorb products from the environment (shaded area in Fig. 2.2, right side). The system produces again the final demand f_i by consuming a direct and indirect amount of (primary) resources from the environment, calculated through relation (2.2).

$$e_{env,i} = \left(r_{env,k} + r_{env,j} + r_{env,i}\right)/f_i \tag{2.2}$$

It is worth to notice that both the inputs and the outputs of every process have to be defined as *one* single kind of product measured with *one* specific unit, and that both physical and monetary units can be adopted (*kWh, kg, units, €* and so on).

Evaluating primary or secondary resources consumed by the simple systems defined by Fig. 2.2 is straightforward. However, modern economies are actually complex networks composed by a very large number of productive processes, connected to each other and with the environment by many different flows of products. This makes the definition of the system to be analyzed a very complex and challenging task. In this perspective, unified rules for the definition of system boundaries and a defined mathematical scheme to account for total resources consumption are required (Finnveden et al. 2009; Guinee et al. 2010; Liao et al. 2012; Suh 2009). Finally, notice that the total resources embodied in one product cannot be defined as a *property* of the defined product, since its value is strictly dependent by the definition of system boundaries.

As stated in Chap. 1, this book focuses on the development of methods to account for primary non-renewable energy-resources embodied in goods and services. However, the accounting methodologies introduced in following sections may be adopted in principle for the evaluation of different types of resources or waste emissions (pollutants, water consumption, soil occupation, material use, etc.).

2.3 Input–Output analysis

Input–Output Analysis (IOA) is the analytical framework originally developed in late 1930s by *Wassily Leontief* in order to analyze and to understand the interdependence of industries within a given economy (Dietzenbacher and Lahr 2004; Leontief 1974). Because of the scientific relevance and the analytical potential of IOA, Leontief was awarded by the Nobel Prize in economics in 1973. Since Leontief's first publications, hundreds of books and articles on input-output analysis have been published. During last decades, original IOA framework have been modified and developed in order to extend its evaluation to other fields, such as: employment and social accounting metrics associated with production activities, regional and interregional flows of products and services, environmental accountings, and so on. Today, IOA is one of the most widely applied methods in both classical economics and in the field Environmental Impact Analysis (Miller and Blair 2009; Suh 2009).

Although IOA is typically applied for the analysis of national economies, its theoretical formulation makes it suitable to describe and to analyze every kind of productive system. After a brief historical overview, the Input-Output analysis is described and formalized as a comprehensive method to account for resources use or waste emissions embodied in products.

2.3.1 Basic Leontief model: single production process

Recalling definitions introduced in paragraph 2.2, Input-Output analysis is here applied to a simple productive system composed only by the generic *i*th process, represented in Fig. 2.3.

Fig. 2.3 Representation of a single production process

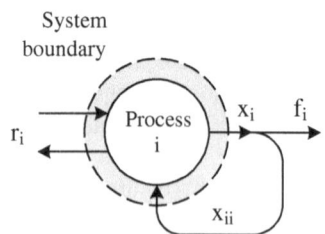

Given a defined time frame, the process produces a net amount of product f_i (the final demand, that is the purpose of system production) consuming a portion of its own product x_{ii} (intermediate transaction), and absorbing a flow of exogenous resources or releasing a flow of wastes r_i (exogenous transactions). The total production of ith process equals the sum of its intermediate consumption and its final demand, as showed by the *production balance* (2.3). The balance have to be written in one homogenous metric, say, *kg, J, units* or *monetary value*.

$$x_i = x_{ii} + f_i \qquad (2.3)$$

For the purpose of IOA, the *technical coefficient* (2.4) is introduced as the ratio between the input to ith process and its total production: it represents the *direct input requirements* to produce one unit of product.

$$a_i = \frac{x_{ii}}{x_i} \qquad (2.4)$$

$$x_i = a_i x_i + f_i \qquad (2.5)$$

By simple algebraic handlings of relation (2.5), total production of ith process can be expressed as a function of its final demand and its technical coefficient: the *Leontief Production Model* (also named *Leontief demand driven model*) is derived as in relation (2.6) (Suh 2005). The *Leontief inverse coefficient* (2.7) (or *Leontief Multiplier*) is the core of the Input-Output analysis: it represents the amount of ith product *embodied* in the unit of the ith final demand.

$$x_i = (1 - a_i)^{-1} f_i \qquad (2.6)$$

$$l_i = (1 - a_i)^{-1} \qquad (2.7)$$

With reference to Fig. 2.3, total production of the ith process (2.6) causes *exogenous transactions of resources or wastes* r_i, such as tons of input materials, Joules of energy, hectares of soil, hours of labour, CO_2 emissions, and so on. In this section, only one single kind exogenous transaction is considered, measured in one single unit. The amount of the considered exogenous transaction embodied in the unit of final demand produced by the ith process e_i can be evaluated thanks to the embodied transaction balance (2.8), which expresses the total transaction of resources or wastes embodied in the whole ith production E_i as the sum of the embodied energy of intermediate inputs E_{ii} plus the exogenous transaction directly absorbed or released by the ith process r_i.

$$E_i = E_{ii} + r_i \rightarrow e_i x_i = e_i x_{ii} + r_i \qquad (2.8)$$

In a similar fashion of technical coefficient (2.4), the *intervention coefficient* (2.9) is defined as the amount of exogenous transaction *directly* required to produce a unit of product.

$$b_i = \frac{r_i}{x_i} \tag{2.9}$$

Introducing the definition of technical coefficient (2.4) in the embodied transaction balance (2.8), and dividing both sides by total production x_i, relation (2.10) is obtained.

$$e_i = e_i a_i + b_i \tag{2.10}$$

The embodied amount of exogenous transaction invoked by the unit of the final demand is then obtained by relation (2.11) and it is expressed as a function of both the technical coefficient and the intervention coefficient of the ith process.

$$e_i = b_i(1 - a_i)^{-1} \tag{2.11}$$

Relation (2.11) is very similar to the Leontief Production Model (2.6): it represents the amount of exogenous transaction *embodied* in the unit of the ith final demand. Although the specific embodied transaction e_i is constant for every unit of product (intermediate production or final demand), it is defined as the amount of exogenous resources needed to produce a unit of *final demand*: indeed, recalling definitions of technical coefficient (2.4) and intervention coefficient (2.9), relation (2.12) is obtained.

$$e_i = \frac{r_i}{x_i} \cdot \frac{x_i}{f_i} = \frac{r_i}{f_i} \tag{2.12}$$

In this simple case, the total exogenous transaction embodied in the ith final demand E_i equals the total amount of exogenous resources consumed by the process r_i, and it results as the product between the specific embodied resources and the final demand, as showed by relation (2.13).

$$E_i = r_i = e_i f_i \rightarrow E_i = r_i = b_i(1 - a_i)^{-1} f_i \tag{2.13}$$

2.3.2 Basic Leontief model: generic system composed by n processes

Let us consider the generic productive system in Fig. 2.4, composed by n productive processes connected to each other by endogenous transactions of goods and services x_{ij} and exogenous transactions of multiple types of resources and wastes r_{ki}. Each process is designed to deliver a certain amount of its product outside system boundary as the final demand f_i. Because of the presence of endogenous interrelations among processes, total production x_i and embodied resources/wastes

Fig. 2.4 Flow of inputs and outputs of a system composed by n production processes

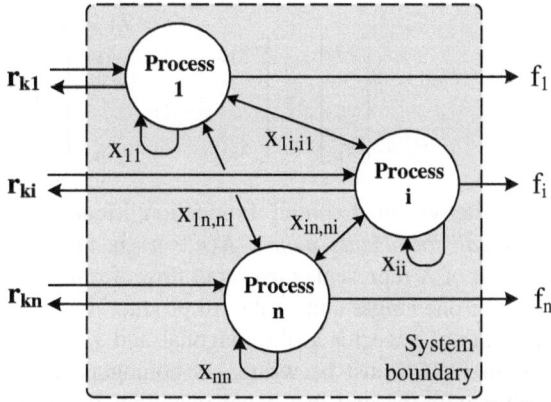

e_i of each single ith productive process is dependent by the other jth processes: as introduced in Sect. 2.2, these values are thus strictly dependent on the definition of the boundaries of the analyzed system.

With reference to Fig. 2.4, it is possible to write one production balance (2.14) for each one of the n productive processes: total production of ith process x_i results as the sum of its self-consumptions x_{ii}, the flows of its products that are required by all the other processes x_{ij}, and the final demand f_i. All the ith balances can be collected in the linear system of equations (2.15).

$$x_i = x_{i1} + \cdots + x_{ij} + \cdots + x_{in} + f_i \rightarrow x_i = \sum_{j=1}^{n} x_{ij} + f_i \qquad (2.14)$$

$$\begin{cases} x_1 = x_{11} + \cdots + x_{1j} + \cdots + x_{1n} + f_1 \\ \qquad \vdots \\ x_i = x_{i1} + \cdots + x_{ij} + \cdots + x_{in} + f_i \\ \qquad \vdots \\ x_n = x_{n1} + \cdots + x_{nj} + \cdots + x_{nn} + f_n \end{cases} \qquad (2.15)$$

System (2.15) can be represented in the compact matrix form (2.16), defining the *total production vector* $\mathbf{x}(n \times 1)$, the *endogenous transaction matrix* $\mathbf{Z}(n \times n)$ (also known as *process-by-process matrix*) and the *final demand vector* $\mathbf{f}(n \times 1)$. In relation (2.16), vector $\mathbf{i}(n \times 1)$ is a column vector of n rows of 1's defined as *summation vector*[1] (Miller and Blair 2009).

---

[1]Generally, to create a column vector whose elements are the row sums of one $(a \times b)$ matrix, it is necessary to post-multiply the considered matrix by the *summation column vector* $\mathbf{i}(b \times 1)$ composed by b rows of 1's. Conversely, pre-multiplication of the same matrix by the *summation row vector* $\mathbf{i}(1 \times a)$ creates a row vector whose elements are the column sums of the $(a \times b)$ matrix. Summation vector $\mathbf{i}$ is useful to express matrix operations in compact form and will be often recalled in the following.

$$\mathbf{x} = \mathbf{Zi} + \mathbf{f}$$

$$\begin{bmatrix} x_1 \\ \vdots \\ x_n \end{bmatrix} = \begin{bmatrix} x_{11} & \cdots & x_{1n} \\ \vdots & \ddots & \vdots \\ x_{n1} & \cdots & x_{nn} \end{bmatrix} \cdot \mathbf{i}\,(n \times 1) + \begin{bmatrix} f_1 \\ \vdots \\ f_n \end{bmatrix} \tag{2.16}$$

To derive the Leontief Production Model (2.6) for this generalized system, *technical coefficients matrix* $\mathbf{A}(n \times n)$ is introduced as in relation (2.17): each element of $\mathbf{A}$ represents the output flows from process ith to process jth required to produce one single unit of the jth product. In (2.17), $\hat{\mathbf{x}}$ is a square matrix with the elements of $\mathbf{x}$ vector at the diagonal and zero elsewhere. Production balances of system (2.16) must be written in homogeneous units, but each balance may be written with the unit that better quantifies its product. In case of multiple units of measure, the technical coefficients matrix $\mathbf{A}$ results in mixed units ($€/€$, $€/kg$, $J/€$, and so on).

$$\mathbf{A}(n \times n) = \mathbf{Z\hat{x}}^{-1} \quad \text{with:} \quad a_{ij} = x_{ij}/x_j$$

$$\begin{bmatrix} a_{11} & \cdots & a_{1n} \\ \vdots & \ddots & \vdots \\ a_{n1} & \cdots & a_{nn} \end{bmatrix} = \begin{bmatrix} x_{11} & \cdots & x_{1n} \\ \vdots & \ddots & \vdots \\ x_{n1} & \cdots & x_{nn} \end{bmatrix} \cdot \begin{bmatrix} 1/x_1 & \cdots & 0 \\ \vdots & \ddots & \vdots \\ 0 & \cdots & 1/x_n \end{bmatrix} \tag{2.17}$$

Thanks to the definition of technical coefficients, system (2.16) can be rewritten in compact form and the *Leontief Production Model* (2.18) is derived for any kind of productive system through simple matrices manipulations. Notice that relation (2.18) allows to account for the total production of each process as a function of technical coefficients matrix $\mathbf{A}$ and final demand vector $\mathbf{f}$.

$$\mathbf{x} = \mathbf{A\hat{x}i} + \mathbf{f} \rightarrow \mathbf{x} = \mathbf{Ax} + \mathbf{f} \rightarrow \mathbf{x} = (\mathbf{I} - \mathbf{A})^{-1}\mathbf{f} \tag{2.18}$$

$$\mathbf{L} = (\mathbf{I} - \mathbf{A})^{-1} \tag{2.19}$$

The *Leontief inverse matrix* $\mathbf{L}$ (2.19) (also named *total requirements matrix* or *multipliers matrix*) is the core of the Input-Output analysis: each element of $\mathbf{L}$ represents the embodied (i.e. *direct* and *indirect*) amount of the ith product required from the jth process in order to deliver one unit of its product as final demand (Miller and Blair 2009).

In a dual way with respect to Sect. 2.3.1, the generic productive system composed by n processes is characterized in terms of m different exogenous transactions of resources and wastes.

The amount of exogenous transactions embodied (i.e. *directly* and *indirectly* invoked) in the system final demand can be evaluated by collecting n embodied exogenous transactions balances as in relation (2.20) into k systems of linear

Eq. (2.21), where the subscript k refers to the kth type of exogenous transaction, e.g. energy, materials, working hours, CO_2 emissions, and so on. For every defined kth kind of exogenous transaction, the ith balance expresses the amount of resource/waste embodied in the ith total production $e_{ki}x_i$ as the sum of all the exogenous transactions embodied in inputs to ith process $e_{kj}x_{ji}$ (self-consumption and all the other intermediate products consumption) and the exogenous transactions directly related with the ith process r_{ki}.

$$e_{ki}x_i = e_{k1}x_{1i} + \cdots + e_{kj}x_{ji} + \cdots + e_{kn}x_{ni} + r_{ki} \qquad \longrightarrow \qquad e_{ki}x_i = \sum_{j=1}^{n} e_{kj}x_{ji} + r_{ki}$$

(2.20)

$$\begin{cases} e_{k1}x_1 = e_{k1}x_{11} + \cdots + e_{ki}x_{i1} + \cdots + e_{kn}x_{n1} + r_{k1} \\ \qquad \vdots \\ e_{ki}x_i = e_{k1}x_{1i} + \cdots + e_{ki}x_{ii} + \cdots + e_{kn}x_{ni} + r_{ki} \\ \qquad \vdots \\ e_{kn}x_n = e_{k1}x_{1n} + \cdots + e_{ki}x_{in} + \cdots + e_{kn}x_{nn} + r_{kn} \end{cases} \quad \begin{array}{l} \forall\,kth \\ \\ exogenous \\ transaction \end{array} \quad (2.21)$$

The exogenous transactions are collected in the *exogenous transactions matrix* $\mathbf{R}(m \times n)$. Every line of matrix $\mathbf{R}$ is expressed in homogeneous units (J, *hours*, *kg*, etc.), and it represents the amount of resources/wastes directly absorbed/released by all the processes. Conversely, every column of matrix $\mathbf{R}$ collects all the different kind of exogenous transactions that are absorbed/released by the ith process. Notice that, depending on the definition of the boundaries of the system, these transactions may be *primary* or *secondary*, according to what introduced in Sect. 2.2.

$$\mathbf{R}(m \times n) = \begin{bmatrix} \mathbf{r_1} \\ {\scriptstyle m \times 1} \end{bmatrix} \cdots \begin{matrix} \mathbf{r_n} \\ {\scriptstyle m \times 1} \end{matrix} \end{bmatrix} \rightarrow \mathbf{R}(m \times n) = \begin{bmatrix} r_{11} & \cdots & r_{1n} \\ \vdots & \ddots & \vdots \\ r_{m1} & \cdots & r_{mn} \end{bmatrix} \begin{array}{l} \text{energy} \\ \text{emissions} \\ \cdots \end{array}$$

(2.22)

To express each kth system (2.21) in matrix form, specific and total embodied exogenous transactions matrices $\mathbf{e}(n \times m)$ and $\mathbf{E}(n \times m)$ are introduced as in relation (2.23), where each element e_{ij} and E_{ij} represents respectively specific and total jth resource/waste directly and indirectly required to produce the ith unit of final demand.

$$\mathbf{e}(n \times m) = \begin{bmatrix} e_{11} & \cdots & e_{1m} \\ \vdots & \ddots & \vdots \\ e_{n1} & \cdots & e_{nm} \end{bmatrix}; \quad \mathbf{E}(n \times m) = \begin{bmatrix} E_{11} & \cdots & E_{1m} \\ \vdots & \ddots & \vdots \\ E_{n1} & \cdots & E_{nm} \end{bmatrix} \quad (2.23)$$

According to the above introduced definitions, each kth system of Eqs. (2.21) can be rewritten in matrix form as in relation (2.24), where superscript "T" refers to the matrix transposition.

$$\hat{x}e = Z^Te + R^T \tag{2.24}$$

Recalling the definition of intervention coefficient (2.9), the *intervention coefficients matrix* $B(m \times n)$ is defined by relation (2.25). Elements of B represent the amount of the kth *direct* exogenous transaction caused by the production of one unit of jth product.

$$B(m \times n) = R\hat{x}^{-1} \quad \text{with:} \quad b_{kj} = r_{kj}/x_j$$

$$
\begin{bmatrix} b_{11} & \cdots & b_{1n} \\ \vdots & \ddots & \vdots \\ b_{m1} & \cdots & b_{mn} \end{bmatrix} = \begin{bmatrix} r_{11} & \cdots & r_{1n} \\ \vdots & \ddots & \vdots \\ r_{m1} & \cdots & r_{mn} \end{bmatrix} \cdot \begin{bmatrix} 1/x_1 & \cdots & 0 \\ \vdots & \ddots & \vdots \\ 0 & \cdots & 1/x_n \end{bmatrix} \tag{2.25}
$$

Based on the above introduced definitions, it is thus possible to rewrite the embodied exogenous transactions balance (2.24) as a function of technical coefficients matrix A and intervention coefficients matrix B, as in relation (2.26).[2]

$$\hat{x}e = \hat{x}A^Te + \hat{x}B^T \rightarrow \hat{x}^{-1}\hat{x}e = \hat{x}^{-1}\hat{x}A^Te + \hat{x}^{-1}\hat{x}B^T \rightarrow e = A^Te + B^T \tag{2.26}$$

The expressions of specific embodied resources in final demand products (2.27) can be derived by handling relation (2.26).

$$Ie - A^Te = B^T \rightarrow e = (I - A^T)^{-1}B^T \rightarrow e = \left[(I - A)^{-1}\right]^T B^T \tag{2.27}$$

Therefore, recalling the definition of Leontief inverse matrix (2.19), specific and total exogenous transactions embodied in final demand can be expressed in compact form by relation (2.28). Notice that elements of total embodied exogenous transactions matrix E obviously differ from the elements of the exogenous transactions matrix R: indeed, matrix E can be naively interpreted as the *allocation* of exogenous transactions among final demand products.

$$e = (BL)^T; \quad E = \hat{f}e \tag{2.28}$$

---

[2]To obtain expression (2.26), the following matrix properties need to be recalled:

- Given two generic matrices $X(n \times m)$ and $Y(m \times r)$, transposition of their product equals the product of the two transposed matrices as follows: $(XY)^T = Y^TX^T$;
- Given an invertible square matrix X, the following identity holds: $XX^{-1} = X^{-1}X = I_n$;
- Given an invertible square matrix X, the following identity holds: $(X^T)^{-1} = (X^{-1})^T$.

Since embodied exogenous transactions is as a *conservative* quantity, the sum of the exogenous resources absorbed by the system equals the sum of the total resources/wastes embodied in its products, as shown by relation (2.29).

$$\left.\begin{array}{l} \mathbf{R}(m \times n) \cdot \mathbf{i}\,(n \times 1) = \mathbf{R_{tot}}(m \times 1) \\ [\mathbf{i}(1 \times n) \cdot \mathbf{E}\,(n \times m)]^{\mathrm{T}} = \mathbf{E_{tot}}(m \times 1) \end{array}\right\} \quad \mathbf{R_{tot}} = \mathbf{E_{tot}} \qquad (2.29)$$

Moreover, if the m different resources or wastes are measured by a unique unit, such as different kind of fossil fuels all measured in *Joules* or different kind of emissions measured in *kg*, relation (2.29) can be further extended and the conservation of total embodied exogenous transactions results as the equality between two scalars, as in relation (2.30).

$$\left.\begin{array}{l} \mathbf{i}(1 \times m) \cdot \mathbf{R_{tot}}(m \times 1) = R_{tot} \\ \mathbf{i}(1 \times m) \cdot \mathbf{E_{tot}}(m \times 1) = E_{tot} \end{array}\right\} \quad R_{tot} = E_{tot} \qquad (2.30)$$

As suggested by *Wiedmann* et al. (Wiedmann et al. 2006), a detailed insight in the structure of embodied resources/wastes for each product of the final demand can be obtained if the Leontief matrix is pre-multiplied by the diagonal matrix of each ith exogenous transaction category, as in relation (2.31), where each jth element of the ith row of both $\underline{\mathbf{e}}(n \times n)$ and $\underline{\mathbf{E}}(n \times n)$ represents direct and indirect contributions of exogenous resources/wastes (respectively specific and total) of the jth process required to deliver the ith final demand.

$$\underline{\mathbf{e}}(n \times n) = \{diag[row_i(\mathbf{B})] \cdot \mathbf{L}\}^{\mathrm{T}} \rightarrow \underline{\mathbf{E}}(n \times n) = \hat{\mathbf{f}} \cdot \underline{\mathbf{e}} \qquad (2.31)$$

Relation (2.31) allows to identify the processes that largely contribute in increasing resources/wastes embodied of each final demand product: indeed, elements in the ith column of $\underline{\mathbf{E}}$ represent the amount of exogenous transactions that are directly invoked by the ith process for its own final demand production, and the exogenous transactions that are directly invoked by the ith process to produce the final demand of all the other processes.

In the following, the Author refers to the *Input-Output table* (IOT) of the considered system as the assembly of endogenous transactions matrix $\mathbf{Z}$, final demand vector $\mathbf{f}$ and total production vector $\mathbf{x}$. Since exogenous transactions matrix $\mathbf{R}$ is also known in literature as the *Environmental Satellite matrix* (Carson 1995; Tukker et al. 2006), the assembly given by IOT and such matrix will be named *Environmentally-Extended Input-Output table*. Moreover, the application of the *Leontief model* refers to the use of relations (2.18) and (2.28) for the evaluation of total production and total resources/wastes invoked for the production of the final demand by a generic system.

2.3.3   Meaning of the Leontief Inverse coefficients

The Leontief inverse matrix **L** is the core of Input-Output analysis: once a system has been defined by means of both technical and intervention coefficients, **L** allows to account for the total exogenous resources/wastes consumption due to the production of a given amount of final demand. The Leontief Inverse coefficient l_i for the ith system composed by one single process (Fig. 2.3) may be directly calculated solving the production balance (2.7); however, it can be alternatively calculated by considering the *power series approximation*, which allows to make clear its economic meaning.

With reference to Fig. 2.5, the production of one unit of final demand by the ith process requires self-consumption of the ith product quantified as a_i, and absorbs a certain amount of exogenous resources b_i. Again, self-consumption of ith product has to be produced in addition to the original unit by the ith process, implying another self-consumption of ith product quantified as a_i^2 and other indirect consumption of $b_i a_i$ exogenous resources. This looping process ends up in the sum of the so called power series visible in relation (2.32).

$$\underbrace{1}_{direct\, requirement} + \underbrace{a_i + a_i^2 + a_i^3 + \cdots}_{indirect\, requirements} = \sum_{t=0}^{+\infty} a_i^t = (1 - a_i)^{-1} \qquad (2.32)$$

Intuitively, it follows that the ith process of Fig. 2.5 can be defined as *productive* only if the net output is positive, that is, only if the power series (2.32) converges to the Leontief inverse coefficient and the necessary and sufficient conditions given in relation (2.33) are respected.

$$x_{ii} < x_i; \quad 0 < a_i < 1; \quad l_i > 1 \qquad (2.33)$$

The concept expressed by power series (2.32) can be extended for the generic production system composed by n processes of Fig. 2.4. The total amount of products and exogenous resources respectively produced and consumed by each ith process to fulfill the final demand of the system results in an infinite series of increasingly smaller contributions: mathematically, the infinite limit of the power

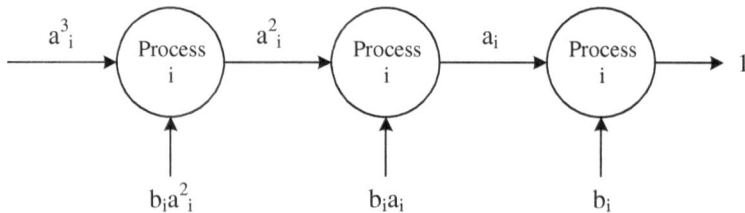

Fig. 2.5 Power series approximation for the single production process

Fig. 2.6 Power series approximation for the system composed by n processes

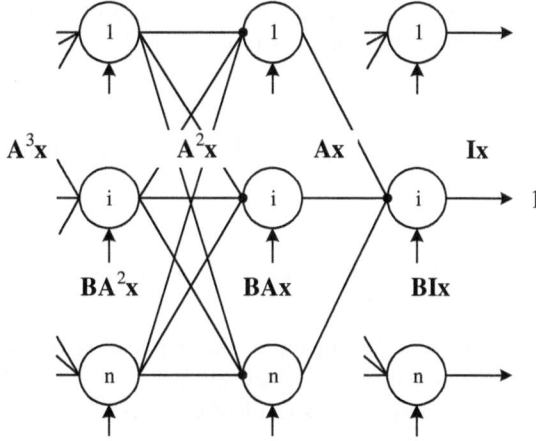

series approaches the Leontief inverse matrix, as suggested by relation (2.34) and graphically showed in Fig. 2.6.

Relation (2.34) reveals its usefulness in a computational perspective, simplifying the matrix inversion process, especially for systems described by dense and large matrices, avoiding numerical problems (Hackbusch 1994; Saad 2003; Watkins 2004).

$$\underbrace{\mathbf{I}}_{\text{direct requirement}} + \underbrace{\mathbf{A} + \mathbf{AA} + \mathbf{AAA} + \cdots}_{\text{indirect requirement}} = \sum_{t=0}^{+\infty} \mathbf{A}^t = (\mathbf{I} - \mathbf{A})^{-1} \qquad (2.34)$$

To assess the convergence of the power series (2.34), that is, to determine whether the system is productive or not, two cases may be distinguished:

- *Homogeneous units.* In case of technical coefficients matrix defined by homogenous units, the sufficient conditions for productivity are: (1) technical coefficients matrix $\mathbf{A}$ is a non-negative matrix and (2) the rows sum of $\mathbf{A}$ results less than one, as in relation (2.35) (Peters 2007; Waugh 1950);

$$a_{ij} > 0 \; \forall i,j \quad \cup \quad \sum_{i=1}^{n} \left| A_{ij} \right| < 1 \; \forall j\text{th columns} \qquad (2.35)$$

- *Non-homogeneous units.* If the entries of the technical coefficients matrix $\mathbf{A}$ are measured in mixed units, conditions given by (2.35) may be no longer respected and it does not give relevant information about system productivity. Therefore, literature indicates the *spectral radius* of **A** as an alternative indicator for power series convergence, as in relation (2.36) (Peters 2007; Varga 2009);

$$\rho(\mathbf{A}) = \{\max(|\lambda|) \ \forall \ \text{eigenvalues of } \mathbf{A}\} < 1 \qquad (2.36)$$

2.3.4 Assumptions of the Leontief Input-Output model

According to the literature, the application of Leontief model to a generic production system is performed through relations (2.18) and (2.28), according to the following assumptions (Ardent et al. 2009; Leontief 1986; Nakamura and Kondo 2009; Treloar 1998):

1. *Process characterization.* As stated in Sect. 2.3.1, every productive process must produce one single kind of product, measured with one specific unit. However, each process may absorb flows of different products from other processes, and different kind exogenous resources from outside system boundaries. As an instance, if one single process produces different kind of chemical reactants, IO table requires these products to be measured in one single unit (e.g. *kg* or *€* of "reactants"). It is not always possible to define production processes with only one product: when different outputs are produced, the embodied resources/wastes caused by the production process needs to be allocated according to one specific criterion (Ekvall and Finnveden 2001);

2. *Technical coefficients.* Leontief model works with *constant returns to scale.* This implies that process technology does not change in a specific time frame. In practice, technical and intervention coefficients are assumed as constant values, resulting in linear production functions: if the output level of a certain process changes, the input requirements (both intermediate products and exogenous resources) will change in a proportional way. This can be considered a reasonable assumption in most cases: given a car production process, if the required final demand of cars increases, steel required by the process will also proportionally increase and vice versa. Obviously, the accuracy of IOA predictions will be lower as the changes in final demand are larger;

3. *Exogenous resource elasticity.* Since the production activity is invoked to meet the final demand, the Leontief model is also known as the *demand driven model* (Miller and Blair 2009). In this perspective, IOA assumes supply of exogenous resources as infinite and perfectly elastic. The demand driven model suits the behavior of human productive systems so far, in which the level of production is driven by the final demand level more than the availability of exogenous resources. When the level of production is limited by the input of exogenous resources, as happens in ecological systems, the so-called *Ghosh* model, or *supply driven model*, can be adopted. Further discussions about *Ghosh* model

are out of the scope of the book and can be retrieved in literature (Dietzenbacher 1989; Ghosh 1958);

4. *Aggregation of processes*. Practical application of IOA for large systems, such as economic systems or supply chains, requires a certain degree of aggregation. For this reason, sometimes it could happen that processes operating in different places are grouped together in one *virtual* sector (e.g. the *steel production sector* of a national economy is formed by all the steel productive plants operating within the considered nation).

2.3.5 Classification of Input-Output Tables

Different kind of Input-Output tables can be found in literature. Commonly, IOTs are classified according to the metric adopted for characterize productive processes: *monetary*, *physical* or *hybrid* units.

Monetary Input-Output Tables (MIOTs). As their name suggest, transaction matrix **A**, final demand vector **f** and total production vector **x** are entirely expressed by means of monetary values. Tables compiled in monetary values are almost exclusively adopted to describe and to analyze national economies, rather than small productive systems. For this reason, the acronym MIOT is here referred always to a monetary IOT of a national economy. MIOTs are compiled according to international rules defined by the *System of National Accounts* (SNA) (Kendrick 1996). MIOTs classify productive activities according to international standards, such as the *International Standard Industrial Classification of all economic activities* (ISIC) (United Nations. Statistical Division 2008) or the *Statistical Classification of Economic Activities in the European Country* (NACE) (Eurostat 2008). In general, MIOTs present all the economic activities being performed for a specific country, pointing out how many goods and services produced by a certain industry in a given year are distributed among the industry itself, other industries, households, etc. The major drawback of MIOTs consists in the approximation of product flows with their monetary equivalents, which makes results affected by variation in products prices. Because of the crucial role covered by MIOTs in the evaluation of the primary exogenous resources/wastes caused by production activities, a deeper discussion about them is given in Chap. 4.

Physical Input-Output Tables (PIOTs). In this IOT, all the entries are measured in physical units, such as mass, energy or number of pieces. *Material Flows Analysis* (MFA) and IOA find a meeting point in PIOTs, that are able to describe material and resource flows within the sectors of the given system (Lifset 2009). Even if PIOTs are usually adopted to analyze small productive systems, these tables were defined also for the following economies: Netherlands, Germany, Denmark, Italy and Finland. Furthermore, a preliminary PIOT for the European Union is based on information from the German and Danish PIOT, scaled up to EU levels. Because of the difficulties in finding required data, these PIOTs are outdated, with lack of information and composed by few sectors only (Hoekstra and van den Bergh 2006). PIOTs suffer of a great number of limitations: firstly, flows are counted in a single unit, usually *tons*: in this way, immaterial products or flows of materials with a high environmental impact cannot be taken into account (usually co-products). Furthermore, a major methodological weakness is related to a not standardized methods for their compilation (Suh and Kagawa 2009; Weisz and Duchin 2006). Finally, compiling PIOTs is time-intensive and the data are not always available for all the economic sectors under consideration. The evidence of this drawbacks is the limited number of PIOTs available in literature.

Recently, several publications performed theoretical comparison among PIOT and MIOT in order to define the best way for the application of Input-Output analysis. The most important difference between PIOT and MIOT is the purpose by which the two tables are designed: PIOT has a purely environmental purpose, whereas MIOT is mainly adopted to have a detailed picture of all economic activities within the economy and to perform various economic performance analysis (Giljum et al. 2004). If both PIOT and MIOT are formulated for the same system with the same number of processes, and if each process sell his products at one unique price, it should be theoretically possible to derive one table from the other one (and vice versa) simply multiplying (or dividing) each entry in the table by its monetary price. In other words, *PIOTs* and *MIOTs are mutually connected by the price of products* (Fisher 1965; Weisz and Duchin 2006). However, this results practically unviable, mainly due to the price inequality of products: every productive sector usually sell his products to other sectors at different prices, thus a matrix of products prices (rather than a vector) need to be identified. Nevertheless, process aggregation makes prices identification a difficult task, introducing large uncertainties in IO models. A deeper discussion about the use of monetary or physical units as metric for IOTs can be found in literature (Giljum et al. 2004; Hubacek and Giljum 2003; Suh 2004; Weisz and Duchin 2006).

Hybrid Input-Output Tables (*HIOTs*). In recent years, some specialists have called for the development of *hybrid* tables with the aim to describe both physical and monetary flows within a given economy: these tables are based on the idea that every sector should have the unit of measurement that best represents the output of that sector (Hourcade et al. 2006). HIOTs requires to build a dual accounting, both in physical and monetary terms, in order to merge these two approaches from the macro-economic point of view. In past decades, construction of HIOTs was constrained because of the limited availability of several monetary and physical data. Nowadays, current data on both physical and economic dimension are mostly available and such account can be performed. In literature, *Konijn* et al. (Konijn et al. 1997) and *Hoekstra* (Hoekstra 2003) have used both physical and monetary units in one IOT, introducing the mixed-unit Input-Output model. Recently, a study about mixed units model for a hybrid energy IOA was proposed by *Mayer* (Mayer and EEA 2007). HIOTs could play an important role in modeling materials and energy flows, solving both the price inhomogeneity typical of MIOTs and the drawback of single-mass unit of PIOTs (Kagawa and Suh 2009). Efforts in HIOTs development are focused on the expansion of the IOA impact evaluation, including different kind of products and able to model social, environmental and economic dimensions of sustainability. For instance, a HIOT was implemented trying to take into account monetary transactions, physical flows of resources and working hours as three potential sustainability indicators (Minx and Baiocchi, 2009).

2.3.6   Data Organization and application of Input-Output analysis

This paragraph provides graphical aided guidelines for the setup of an *Environmentally Extended Input-Output table* of a generic productive system, the application of the Leontief Model and a discussion about the analytical potential of Input-Output analysis.

Application guidelines

Practical transposition of the LCA guidelines listed in Sect. 2.1 are here proposed for the application of Input-Output analysis. Recalling the nomenclature presented in Sect. 2.2, the following application steps are proposed:

1. *Objective and Functional Unit.* It consists in the definition of the objective of the Input-Output analysis, that is, the product or service produced by the system as the final demand (i.e. the functional unit). Moreover, also the kind of exogenous transactions to be evaluated must be defined. For instance, the objective of IOA could be the evaluation of electric energy required by a firm to produce one specific plastic bag, or the evaluation of primary fossil fuels required by each sector of a national economy to produce its products;
2. *Temporal and spatial boundaries.* This second step consists in the definition of temporal and spatial extensions of the considered system. Different life cycle phases (production, use, disposal) and spatial boundaries can be arbitrarily defined, and both these choices largely affects the final results. Notice that the effects due to all the processes that lies outside time and space boundaries of the system will be neglected: in the example of the plastic bag, setting boundaries of IOA as the physical boundaries of the factory will cause the exclusion from the exogenous transactions accounting of all the upstream electric energy requirements;
3. *Process characterization.* In this phase, every productive process has to be defined and characterized in terms of goods and services production and exogenous resources/wastes transactions, and the two following conditions have to be met: (1) each production process must produce only one kind of product, measured with one single metric and (2) every production process must delivers inputs to, or receives outputs from, other processes. This phase leads to the definition of technical coefficients matrix **A** and intervention coefficients matrix **B**;

4. *Application of Leontief model and results interpretation.* In this phase, IOTs are compiled and Leontief model is applied according to relations (2.18) and (2.28), evaluating the total production $\mathbf{x}$ required to sustain a given level of final demand $\mathbf{f}$, and the related specific and total embodied exogenous transactions $\mathbf{e}$ and $\mathbf{E}$;

5. *Uncertainty analysis.* Input data to IOA could be affected by uncertainties that propagate through the Leontief model till final results: for this reason, uncertainty analysis is relevant. This issue has been addressed in LCA by many authors (Heijungs 1994, 1996; Heijungs and Huijbregts 2004). The widely adopted approach has been developed under the name of *Marginal Analysis* or *Perturbation Analysis* (Heijungs and Suh 2002a): uncertainties of results by IOA can be measured by introducing numerical perturbations in technical coefficients and input matrices, determining how such perturbations affect the demand of exogenous resources or the production of wastes. Finally, statistical treatment and post-processing of the obtained results should be performed.

Further methodological details about the application of Input-Output analysis can be retrieved in literature (Suh and Huppes 2005).

Environmentally Extended Input-Output Table and application of Leontief model

For the purpose of practical IOA applications, let us consider the generic system depicted in Fig. 2.4, composed by n productive processes, producing n different products and exchanging m different kind of flows with the environment. With reference to Fig. 2.7, rows of the endogenous transactions matrix $\mathbf{Z}$ represents products of the ith process received as the inputs to jth process, elements in final demand vector f represent the net useful output of each ith process, and elements of the exogenous transactions matrix $\mathbf{R}$ represent the flows of resources or wastes directed to all the processes that cross system boundaries. Practical application of the Leontief model, expressed by means of relations (2.18) and (2.28), requires all

From / To	1 ... n	Final demand	Total production
Process 1 ... Process n	Endogenous transactions matrix $\mathbf{Z}(n \times n)$	$\mathbf{f}(n \times 1)$	$\mathbf{x}(n \times 1)$
Resource 1 ... Resource m	Exogenous transactions matrix $\mathbf{R}(m \times n)$		

Fig. 2.7 Input Output table for generic n-process system augmented with exogenous resources matrix

the above introduced matrices to be respectively converted into the technical coefficients matrix **A** and intervention coefficients matrix **B**, as showed by Fig. 2.8. Finally, Fig. 2.9 reports the essential relations required for the derivation of specific and total exogenous transactions embodied in the final demand.

The analytical potential of Input-Output analysis

The main two useful applications of Leontief model are here listed and described. A deeper theoretical discussion with applications can be retrieved in literature (Nakamura and Kondo 2009).

- *Effects of final demand shocks.* Given the Environmentally Extended Input-Output table of a generic system in the form of Fig. 2.7, application of

From / To	1 ... n	Final demand	Total production
Process 1 ... Process n	Technical coefficients matrix $A(n \times n) = Z\hat{x}^{-1}$	$f(n \times 1)$	$x(n \times 1)$
Resource 1 ... Resource m	Intervention coefficients matrix $B(m \times n) = R\hat{x}^{-1}$		

Fig. 2.8 Technical coefficients and input matrices for generic n-process system

From / To	Processes 1 ... n		Resources 1 ... m	Resources 1 ... m
Process 1 ... Process n	Leontief inverse matrix $L = (I - A)^{-1}$		Specific embodied exogenous transactions matrix $e = (BL)^{T}$	Total embodied exogenous transactions matrix $E = \hat{f}e$
Total				Total embodied exogenous transactions vector $E_{tot} = i(1 \times n)E$

Fig. 2.9 Application of Leontief model

Leontief model as in relation (2.37) reveals how a change in the final demand of the system respectively affects its total production and total embodied exogenous transactions;

$$\Delta \mathbf{x} = \mathbf{L} \Delta \mathbf{f}; \quad \Delta \mathbf{E} = \Delta \hat{\mathbf{f}} (\mathbf{BL})^{\mathrm{T}} \tag{2.37}$$

- *Effects of technological shocks.* Changes in technology of one or more productive processes of a same system result in a change in technical coefficients matrix from $\mathbf{A}$ to $\bar{\mathbf{A}}$, and input coefficients matrix from $\mathbf{B}$ to $\bar{\mathbf{B}}$. As a consequence, Leontief inverse coefficients matrix will also changes from $\mathbf{L}$ to $\bar{\mathbf{L}}$. Applying Leontief model to this new system configuration by keeping constant the final demand, as in relations (2.38) and (2.39), allows to estimate the effects that a change in technology may have in terms of total production and exogenous transactions embodied in the final demand;

$$\Delta \mathbf{x} = (\mathbf{L} - \bar{\mathbf{L}}) \mathbf{f} \tag{2.38}$$

$$\Delta \mathbf{e} = (\mathbf{BL})^{\mathrm{T}} - (\overline{\mathbf{BL}})^{\mathrm{T}} \rightarrow \Delta \mathbf{E} = \hat{\mathbf{f}} \left[(\overline{\mathbf{BL}})^{\mathrm{T}} - (\overline{\mathbf{BL}})^{\mathrm{T}} \right] \tag{2.39}$$

- *Structural Decomposition Analysis (SDA).* Usually, structural changes in system production affect both the technology and the level of final demand production at the same time. Therefore, it is important to decouple these two overlapped effects and to account for them separately. This is performed in literature through the so-called *Structural Decomposition Analysis* (SDA), formalized here through relation (2.40) and applied to one system defined by two different technological and productive conditions (subscripts 0 and 1) (Nakamura and Kondo 2009). Many applications of SDA to the analysis of environmental burdens can be found in literature (Dietzenbacher and Los 1998; Hoekstra and Van Den Bergh 2002; Kagawa and Inamura 2001).

$$\Delta \mathbf{E} = \Delta \mathbf{E_T} + \Delta \mathbf{E_F} = \underbrace{\hat{\mathbf{f}}_1 \cdot (\mathbf{e}_1 - \mathbf{e}_0)}_{T-technology} + \underbrace{\left(\hat{\mathbf{f}}_1 - \hat{\mathbf{f}}_0 \right) \cdot \mathbf{e}_0}_{F-final\ demand} \tag{2.40}$$

2.3.7 Practical application of Input-Output analysis

According to the application guidelines defined in Sect. 2.3.6, IOA is here practically applied to a simple and ideal factory that produces *plastic bags*, composed by three processes: *Bags production* (1), *Plastic production* (2) and *Heat production* (3), outlined in Fig. 2.10.

Fig. 2.10 Plastic bags
production system

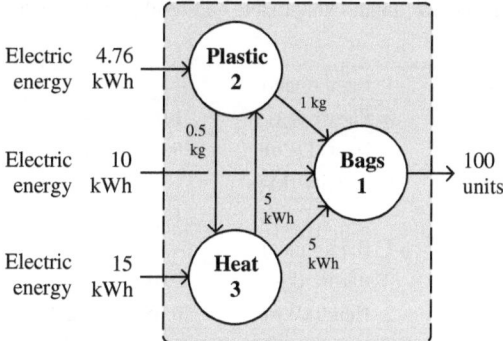

Objective and Functional Unit. The objective of the Input-Output analysis is to account for the amount of electricity embodied in 100 plastic bags produced by the process 1 in Fig. 2.10. In this case, the Functional Unit consists in 100 plastic bags.

Temporal and spatial boundaries. The analysis focuses on the evaluation of the embodied exogenous resources invoked by the system. Therefore, the time required to produce 100 plastic bags is assumed as the time boundary for the analysis. Spatial boundaries are limited to the physical extension of the factory, encompassing the three productive processes outlined in Fig. 2.10. Other life cycle phases, as well as the other upstream/downstream processes out of the considered boundaries, are thus excluded from the analysis.

Process characterization. Every process produces one single output measured in one single unit, receiving one or more inputs from other processes and from outside the system. Notice that electric energy is the only exogenous resource and it feed all the three processes. Values reported in Fig. 2.10 can be arranged in an Physical Input-Output table, defining endogenous and exogenous transactions matrices $\mathbf{Z}$ and $\mathbf{R}$, final demand and total production vectors $\mathbf{f}$ and $\mathbf{x}$. Notice that in this simple example the exogenous transactions matrix $\mathbf{R}$ consists in a vector, since only one kind of exogenous resource is taken into account.

Application of Leontief model and results interpretation. The problem can be mathematically formalized as in Table 2.1. Once technical and intervention coefficients matrices $\mathbf{A}$ and $\mathbf{B}$ have been compiled, the Leontief model can be applied by means of relations (2.18) and (2.28). Results of the analysis are showed in Table 2.1: the specific and total values of electricity embodied in the functional unit result respectively as 0.32 kWh/unit and 32 kWh. It is worth to notice that the latter is three times greater than the direct electricity consumption (10 kWh) caused by the plastic bag production process. Since the final demand is formed by plastic bags only, values of electricity embodied in units of the other products do not have a practical meaning: all the exogenous electricity requirements will be charged on plastic bags only. Because rows sum of the total embodied exogenous resources vector equals the column sum of the exogenous resources (32 kWh of electricity), identity (2.30) is respected and results of IOA are correct.

Table 2.1 Input-Output table and results of the application of Leontief model

		1	2	3	f	x
Z	1. Bags (units)	–	–	–	100	100
	2. Plastic (kg)	1.0	–	0.5	0	1.5
	3. Heat (kWh)	5.0	5.0	–	0	10
R	Electricity (kWh)	10	7	15		
		1	2	3	f	x
A	1. Bags (units)	–	–	–	100	100
	2. Plastic (kg)	0.01	–	0.05	0	1.5
	3. Heat (kWh)	0.05	3.33	–	0	10
B	Electricity (kWh)	0.10	4.67	1.50		
		1	2	3	e (kWh/…)	E (kWh)
L	1. Bags (units)	–	–	–	0.3	32
	2. Plastic (kg)	0.00	–	0.01	11.6	0
	3. Heat (kWh)	0.00	2.22	–	2.1	0
Total						32

Due to the simplicity of the given example, accounting for electricity embodied in the final products is straightforward. Advantages in the application of Input-Output analysis thus appear to be relevant in case of more complex systems, composed by a large number of components with multiple kind of endogenous and exogenous transactions of products, resources and wastes.

2.4 Process analysis

One alternative method to account for resources or wastes embodied in one given good or service is the so-called *Process analysis* (PA). This method was defined for the first time by the *International Federation of Institutes for Advanced Study* (IFIAS) in 1974 in order to define a standard method to account for primary energy embodied in products, overcoming the problem due to the lack of supply chains data (International Federation of Institutes for Advanced Studies 1978). Once the objective of the analysis (i.e. the functional unit) has been defined, evaluation of the primary energy embodied in products through PA consists in tracing back the upstream structure of its supply chains according to scheme depicted in Fig. 2.11, identifying every upstream contribution in terms of primary energy directly taken from the environment level by level.

Detailed definition and application of PA for the evaluation of embodied energy of products has been carried out by *Trelorar* (Treloar 1998) and *Wilting* (Wilting 1996). These studies reveal that complexity of the network and data requirements increase exponentially as the levels of the analysis increase but, at the same time, primary

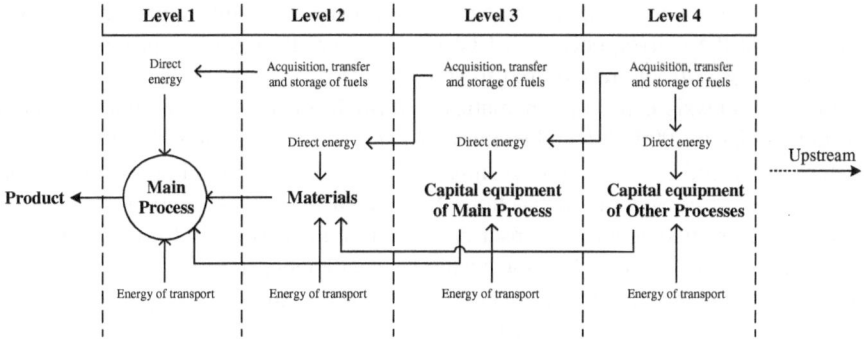

Fig. 2.11 Standard procedure to account for primary energy of products through process analysis

energy requirements decrease level by level until reaching negligible contributions on the total embodied energy of the analyzed product (Chapman et al. 1974).

During the last decades, many researchers and organizations have worked to develop specific process-based models for LCA analysis (Curran 1996; In and Curran 1994; Keoleian et al. 1993) aimed at evaluating embodied resources of specific products and systems. For this purpose, different commercial software have been developed, such as *Simapro*® (Goedkoop et al. 2004; PRé 2008) or *GaBi*® (Spatari et al. 2001), and process-based models were collected in commercial databases, such as *Ecoinvent* (Frischknecht et al. 2005). Except for few methodological discussions (Consoli et al. 1993; Hendrickson et al. 1997; Suh and Heijungs 2007; Suh and Huppes 2005; Tillman et al. 1994), agreement about the best theoretical formalization of PA as a primary resource accounting method is still nonexistent.

2.4.1 Definition and formalization of Process analysis

In order to apply Process analysis to a generic system, a practical transposition of the LCA guidelines listed in Sect. 2.1 is here proposed. These guidelines are very similar to the ones proposed for the application of Input-Output analysis in Sect. 2.3.6: (1) *Objective and Functional Unit definition*, (2) *Temporal and spatial boundaries definition*, (3) *Process characterization* and (4) *Calculation of embodied transactions and results interpretation*. The essential element in the application of the above introduced procedure is the so-called *process flow diagram*: it consists in a process tree similar to the one presented in Fig. 2.11. Once the process tree have been characterized in terms of endogenous and exogenous transactions, the total resources/wastes embodied in the considered product can be calculated. As will be practically showed, and as underlined by *Consoli* et al., this calculation process results in an infinite geometric progression if the considered system presents *internal loops of intermediate transactions*: in this case, an iterative approach is required (Consoli et al. 1993). As the levels of the process flow diagram

increase, the number of required iterations also increase: results approach the exact
solution and the convergence speed become slower. Therefore, appropriate selection of *cut-off criteria* are required to stop the calculations.

Process analysis can be mathematically formalized by means of linear algebra
(Miller and Blair 2009; Suh and Heijungs 2007). The application of PA to a generic
system composed by n processes results in the calculation of total production $\mathbf{x}$ and
total embodied exogenous transactions $\mathbf{E}$ invoked for the production of the final
demand $\mathbf{f}$, classified as direct (subscript D) and indirect (subscript I) contributions.
In relations (2.41) and (2.42), the technical coefficients matrix $\mathbf{A}$ and the intervention coefficients matrix $\mathbf{B}$ have been defined in the same way as for
the Input-Output analysis (see Sect. 2.3.2).

$$\mathbf{x} = \mathbf{x}_D + \mathbf{x}_I \rightarrow \mathbf{x} = \mathbf{I} \cdot \mathbf{f} + \mathbf{A} \cdot \mathbf{f} + \mathbf{AA} \cdot \mathbf{f} + \mathbf{AAA} \cdot \mathbf{f} + \cdots \qquad (2.41)$$

$$E = E_D + E_I \rightarrow \mathbf{E} = \mathbf{BIf} + \mathbf{B}(\mathbf{Af} + \mathbf{AAf} + \mathbf{AAAf} + \cdots) \qquad (2.42)$$

As can be inferred by Eqs. (2.41) and (2.42), results of Process analysis con-
verge to the results of Input-Output analysis through the *power series approximation* (2.34): this reveals a strong theoretical relation between Process Analysis
and Input-Output analysis.

2.4.2   Practical application of Process analysis

In this paragraph, PA is applied to the simple system outlined in Fig. 2.4. The
application of PA has the same objective, temporal and spatial extensions defined
for the application of IOA.

Figure 2.12 resumes the phase of characterization of processes required for the
application of PA: in that phase, specific endogenous and exogenous inputs to each
process are evaluated. This representation is useful in order to draw the process flow
diagram as showed in Fig. 2.13, in which the amount of products exchanged by all
the processes are visible in the *tree structure* typical of process-based models. With
reference to Fig. 2.13, the direct electricity requirements of Bags process are
already known from the data provided by the process characterization phase:

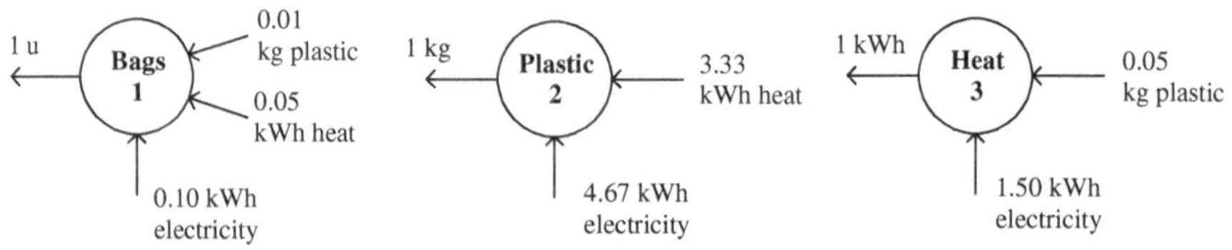

Fig. 2.12 Characterization of processes for the application of PA

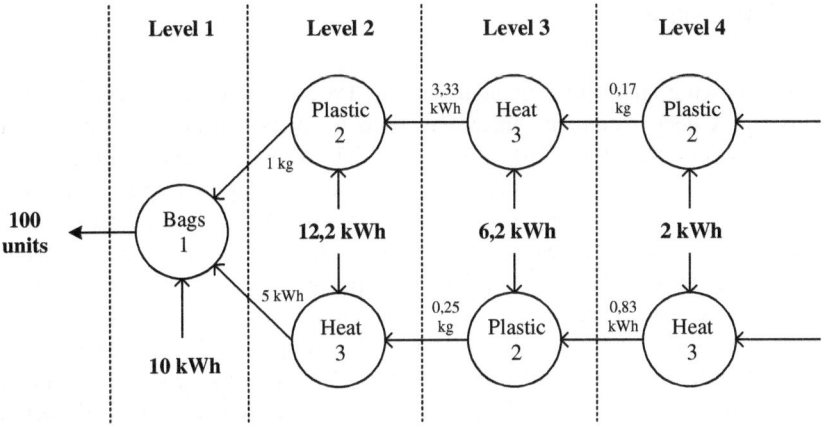

Fig. 2.13 Direct and Indirect contributions to total production and total embodied exogenous transactions

indeed, the Bags process produces 1 bag by absorbing 0.1 kWh, and thus resulting in 10 kWh of total direct electricity requirements. To assess the indirect contributions to total electricity consumption caused by production of bags, further levels of the process flow diagram are considered according to relations (2.41) and (2.42).

Results of Process analysis are reported in Table 2.2: indirect contributions to embodied exogenous transactions of electricity have been traced till level 4 (see Fig. 2.13). The analysis is complete since the numerical difference of indirect contributions between levels 3 and 4 are very small, thus the electricity embodied in 100 bags turns out to be 31.4 kWh. However, the latter could not always be adopted as the only convergence criterion: it has been demonstrated that, especially for the analysis of large systems, diminishing the amount of exogenous resources required for each level provides no guarantee that the sum of that single negligible contributions is also negligible (Bullard et al. 1978).

Table 2.2 Results of process analysis

	n	Name	Units	0	1	2	3	4
Processes	1	Bags	u	100	–	–	–	–
	2	Plastic	kg	–	1.00	0.25	0.17	0.04
	3	Heat	kWh	–	5.00	3.33	0.83	0.56
Electricity consumption	R		kWh	10.0	12.2	6.2	2.0	1.0
Embodied electricity	E		kWh	10.0	22.2	28.3	30.4	31.4

2.5   Discussion

In the following, Input-Output analysis and Process analysis are compared and discussed. A deeper comparison among such methods can be found in literature (Hendrickson et al. 1997, 2010; Miller and Blair 2009; Suh and Heijungs 2007).

Mathematical equivalency between IOA and PA. Input-Output analysis and Process analysis often emerges from the literature as two distinct techniques to account for embodied resources/wastes in products. Generally, PA is used to account for the primary resources required in specific productive processes, whereas IOA is mostly used for economic and environmental impact analysis of productive sectors of national economies. The theoretical description of PA and IOA models presented in this chapter reveals that once the objective of the analysis (i.e. the final demand) and the boundaries of the productive system have been univocally defined, *application of PA converges to IOA trough the power series approximation* (2.34) (Suh and Huppes 2005). *It follows that any productive system can be always described and analyzed by means of IOA, which is a better formalized method with respect to PA.* However, if the analyzed system is very simple (formed by few productive processes) and it does not present any loop of endogenous transactions, PA turns out to be simpler than IOA.

Boundaries definition. Every application of both PA and IOA requires to define the boundaries of the system, and to collect data from scratch. This makes results of the analysis strongly dependent by the analyst's choices, making two different analyses of a same product not comparable. This is particularly relevant in the evaluation of *primary* resources/wastes embodied in products, for which the analyzed system can be characterized by a very large number of processes. Moreover, definition of process trees similar to the one in Fig. 2.11 call to focus on relevant flows of products or services, disregarding small contributions that could actually be relevant in the evaluation of the total amount of embodied resources/wastes. Indeed, literature shows that the environmental burdens linked to negligible inputs could be non-negligible (Hendrickson et al. 1997; Lave 1995).

Accounting for the effects of immaterial services and human labour. Both PA and IOA are typically focused on productive processes with material outputs. Therefore, environmental effects caused by production of immaterial services are usually neglected (even in methodological discussions) mainly because of difficulties in their characterization, making the embodied resources/wastes accounting not fully comprehensive (Treloar 1998). Moreover, the role of human labour in resources consumption is a controversial and largely debated topic in LCA: in principle, production of one working hour required by one specific productive process requires consumption of goods and services in turn, causing additional environmental impact. Literature states that such contributions would result negligible (Boustead and Hancock 1979) but a numerical proof of this statement has not been provided yet.

Based on the theoretical discussion of this chapter, and with reference to the emerging needs of environmental impact analysis highlighted in Sect. 1.2,

Input-Output analysis will be adopted in this book as the reference method to account for embodied resources/wastes in products. More specifically, IOA will be adopted to: (1) account for *primary* energy-resources embodied in detailed products and services in a reproducible, standardized and simple way; (2) account for the overall thermodynamic performances of energy conversion systems in their whole life cycle; (3) internalize the effects of human labor in environmental accountings.

References

Ardent, F., Beccali, M., & Cellura, M., (2009). Application of the IO methodology to the energy and environmental analysis of a regional context. In *Handbook of input-output economics in industrial ecology* (pp. 435–457). Berlin: Springer.

Meadows D. H., Meadows D. L., Randers J., Behrens, W. W. (1972). The limits to growth. In *A Report for The Club of Rome's Project on the Predicament of Mankind*.

Boustead, I., & Hancock, G. F. (1979). *Handbook of industrial energy analysis*. New York: Wiley.

Bullard, C. W, I. I. I., & Herendeen, R. A. (1975). The energy cost of goods and services. *Energy Policy, 3*, 268–278.

Bullard, C. W., Penner, P. S., & Pilati, D. A. (1978). Net energy analysis: Handbook for combining process and input-output analysis. *Resources and Energy, 1*, 267–313.

Carson, C. S. (1995). Integrated economic and environmental satellite accounts. *Nonrenewable Resources, 4*, 12–33.

Chapman, P. (1974). Energy costs: A review of methods. *Energy Policy, 2*, 91–103.

Chapman, P. F., & Faculty, S. (1975). The energy costs of materials. *Energy Policy, 3*, 47–57.

Chapman, P., Leach, G., & Slesser, M. (1974). The energy cost of fuels. *Energy Policy, 2*, 231–243.

Consoli, F., Allen, D., Boustead, I., Fava, J., Franklin, W., & Jensen, A., et al. (1993). Guidelines for life-cycle assessment: A 'code of practice'. In *Society of environmental toxicology and chemistry (SETAC)*, Pensacola, FL, USA.

Costanza, R. (1980). Embodied energy and economic valuation. *Science, 210*, 1219–1224.

Costanza, R. (2004). Value theory and energy. *Encyclopedia of Energy, 6*, 337–346.

Curran, M. A. (1996). Environmental life-cycle assessment. *International Journal of Life Cycle Assessment, 1*, 179.

Dietzenbacher, E. (1989). On the relationship between the supply-driven and the demand-driven input-output model. *Environment and Planning A, 21*, 1533–1539.

Dietzenbacher, E., & Lahr, M. L. (2004). *Wassily Leontief and input-output economics*. Cambridge: Cambridge University Press.

Dietzenbacher, E., & Los, B. (1998). Structural decomposition techniques: sense and sensitivity. *Economic Systems Research, 10*, 307–324.

Dixit, M. K., Fernández-Solís, J. L., Lavy, S., & Culp, C. H. (2010). Identification of parameters for embodied energy measurement: A literature review. *Energy and Buildings, 42*, 1238–1247.

Duchin, F. (1992). Industrial input-output analysis: Implications for industrial ecology. *Proceedings of the National Academy of Sciences, 89*, 851–855.

Duchin, F. (1998). *Structural economics: Measuring change in technology, lifestyles, and the environment*. Island Press.

Duchin, F., Lange, G.-M., Thonstad, K., Idenburg, A., & Cropper, M. L. (1996). The future of the environment: Ecological economics and technological change. *Journal of Economic Literature, 34*, 818–819.

Ekvall, T., & Finnveden, G. (2001). Allocation in ISO 14041—A critical review. *Journal of Cleaner Production, 9*, 197–208.

Eurostat. (2008). NACE rev.2. *Statistical classification of economic activities in the European Community.* Luxembourg: Office for Official Publications of. the European Communities.

Fay, R., Treloar, G., & Iyer-Raniga, U. (2000). Life-cycle energy analysis of buildings: A case study. *Building Research & Information, 28*, 31–41.

Finnveden, G., Hauschild, M. Z., Ekvall, T., Guinée, J., Heijungs, R., Hellweg, S., et al. (2009). Recent developments in life cycle assessment. *Journal of Environmental Management, 91*, 1–21.

Fisher, F. M. (1965). Choice of units, column sums, and stability in linear dynamic systems with nonnegative square matrices. *Econometrica: Journal of the Econometric Society, 445–450.*

Frischknecht, R., Jungbluth, N., Althaus, H.-J., Doka, G., Dones, R., Heck, T., et al. (2005). The ecoinvent database: Overview and methodological framework (7 pp). *International Journal of Life Cycle Assessment, 10*, 3–9.

Ghosh, A. (1958). Input-output approach in an allocation system. *Economica, 58–64.*

Giljum, S., Hubacek, K., & Sun, L. (2004). Beyond the simple material balance: A reply to Sangwon Suh's note on physical input–output analysis. *Ecological Economics, 48*, 19–22.

Goedkoop, M., Oele, M., & Effting, S. (2004). *SimaPro database manual methods library.* Netherlands: PRé Consultants.

Guinée, J. B. (2002a). Handbook on life cycle assessment operational guide to the ISO standards. *International Journal of Life Cycle Assessment, 7*, 311–313.

Guinée, J. B. (2002b). Handbook on life cycle assessment operational guide to the ISO standards. *The International Journal of Life Cycle Assessment, 7*, 311–313.

Guinee, J. B., Heijungs, R., Huppes, G., Zamagni, A., Masoni, P., Buonamici, R., et al. (2010). Life cycle assessment: Past, present, and future†. *Environmental Science and Technology, 45*, 90–96.

Hackbusch, W. (1994). *Iterative solution of large sparse systems of equations.* Berlin: Springer.

Hammond, G. P., & Jones, C. I. (2008). Embodied energy and carbon in construction materials. *Proceedings of the Institution of Civil Engineers-Energy, 161*, 87–98.

Heijungs, R. (1994). A generic method for the identification of options for cleaner products. *Ecological Economics, 10*, 69–81.

Heijungs, R. (1996). Identification of key issues for further investigation in improving the reliability of life-cycle assessments. *Journal of Cleaner Production, 4*, 159–166.

Heijungs, R., & Huijbregts, M. A. (2004). A review of approaches to treat uncertainty in LCA. In *Proceedings of the IEMSS conference*, Osnabruck.

Heijungs, R., & Suh, S. (2002a). *The computational structure of life cycle assessment* (Vol. 11).

Heijungs, R., & Suh, S. (2002b). *The computational structure of life cycle assessment.* Berlin: Springer.

Hendrickson, C. T., Horvath, A., Joshi, S., Klausner, M., Lave, L. B., & McMichael, F. C. (1997a). Comparing two life cycle assessment approaches: A process model vs. economic input-output-based assessment. In *Proceedings of the 1997 IEEE International symposium on electronics and the environment, 1997* (pp. 176–181). ISEE-1997.

Hendrickson, C. T., Lave, L. B., & Matthews, H. S. (2010). *Environmental life cycle assessment of goods and services: An input-output approach.* Routledge.

Herendeen, R., & Tanaka, J. (1976). Energy cost of living. *Energy, 1*, 165–178.

Hoekstra, R. (2003). *Structural change of the physical economy. Decomposition analysis of physical and hybrid-unit input-output tables.*

Hoekstra, R., & Van Den Bergh, J. C. J. M. (2002). Structural decomposition analysis of physical flows in the economy. *Environmental & Resource Economics, 23*, 357–378.

Hoekstra, R., & van den Bergh, J. C. (2006). Constructing physical input–output tables for environmental modeling and accounting: Framework and illustrations. *Ecological Economics, 59*, 375–393.

Hourcade, J.-C., Jaccard, M., Bataille, C., & Ghersi, F. (2006). Hybrid modeling: New answers to old challenges introduction to the special issue of "The Energy Journal". *The Energy Journal,* 1–11.

Hubacek, K., & Giljum, S. (2003). Applying physical input–output analysis to estimate land appropriation (ecological footprints) of international trade activities. *Ecological Economics, 44*, 137–151.

In, B. M., Curran, M. A. (1994). *Life-cycle assessment: Inventory guidelines and principles*. CRC Press.

International Federation of Institutes for Advanced Studies, I. (1978). IFIAS workshop report, energy analysis and economics. *Resources and Energy, 1*, 151–204.

Joshi, S. (1999). Product environmental life-cycle assessment using input-output techniques. *Journal of Industrial Ecology, 3*, 95–120.

Kagawa, S., & Inamura, H. (2001). A structural decomposition of energy consumption based on a hybrid rectangular input-output framework: Japan's case. *Economic Systems Research, 13*, 339–363.

Kagawa, S., & Suh, S. (2009). Multistage process-based make-use system. *Handbook of Input-Output Economics in Industrial Ecology*, 777–800 (Springer).

Kendrick, J. W. (1996). *The new system of national accounts*. Berlin: Springer.

Keoleian, G. A., Menerey, D., & Curran, M. (1993). *Life cycle design: Guidance manual*.

Klöpffer, W. (1997). Life cycle assessment. *Environmental Science and Pollution Research, 4*, 223–228.

Konijn, P., de Boer, S., & van Dalen, J. (1997). Input-output analysis of material flows with application to iron, steel and zinc. *Structural Change and Economic Dynamics, 8*, 129–153.

Lave, L. B. (1995). Using input-output analysis to estimate economy-wide discharges. *Environmental Science and Technology, 29*, 420A–426A.

Lave, L., MacLean, H., Hendrickson, C., & Lankey, R. (2000). Life-cycle analysis of alternative automobile fuel/propulsion technologies. *Environmental Science and Technology, 34*, 3598–3605.

Lenzen, M., & Dey, C. (2000). Truncation error in embodied energy analyses of basic iron and steel products. *Energy, 25*, 577–585.

Lenzen, M., & Munksgaard, J. (2002). Energy and CO_2 life-cycle analyses of wind turbines—Review and applications. *Renewable Energy, 26*, 339–362.

Leontief, W. (1974). Structure of the world economy: Outline of a simple input-output formulation. *The American Economic Review*, 823–834.

Leontief, W. (1986). *Input-output economics*. Oxford: Oxford University Press.

Liao, W., Heijungs, R., & Huppes, G. (2012). Thermodynamic analysis of human–environment systems: A review focused on industrial ecology. *Ecological Modelling, 228*, 76–88.

Lifset, R. (2009). Industrial ecology in the age of input-output analysis. *Handbook of Input-Output Economics in Industrial Ecology*, 3–21 (Springer).

Mayer, H., & EEA. (2007). Calculation and analysis of a hybrid energy input-output table for Germany within the environmental-economic accounting (EEA). In *The 16th International input-output conference* (pp. 2–6).

Miller, R. E., Blair, P. D. (2009). Input-output analysis: Foundations and extensions. Cambridge: Cambridge University Press.

Minx, J. C., Baiocchi, G. (2009). Time use and sustainability: An input-output approach in mixed units. *Handbook of Input-Output Economics in Industrial Ecology*, 819–846 (Springer).

Nakamura, S., & Kondo, Y. (2009). Waste input-output analysis: Concepts and application to industrial ecology. Berlin: Springer.

Nakamura, S., Nakajima, K., Kondo, Y., & Nagasaka, T. (2007). The waste input-output approach to materials flow analysis. *Journal of Industrial Ecology, 11*, 50–63.

Pennington, D., Potting, J., Finnveden, G., Lindeijer, E., Jolliet, O., Rydberg, T., et al. (2004). Life cycle assessment Part 2: Current impact assessment practice. *Environment International, 30*, 721–739.

Peters, G. P. (2007). Efficient algorithms for life cycle assessment, input-output analysis, and Monte-Carlo analysis. *International Journal of Life Cycle Assessment, 12*, 373–380.

Pokrovskii, V. N. (2011). Econodynamics: The theory of social production. Berlin: Springer.

PRé. (2008). SimaPro LCA software. *Amersfoort, Netherlands: PRé product ecology consultants.*

Reap, J., Roman, F., Duncan, S., & Bras, B. (2008). A survey of unresolved problems in life cycle assessment. *International Journal of Life Cycle Assessment, 13*, 374–388.

Rebitzer, G., Ekvall, T., Frischknecht, R., Hunkeler, D., Norris, G., Rydberg, T., et al. (2004). Life cycle assessment Part 1: Framework, goal and scope definition, inventory analysis, and applications. *Environment International, 30*, 701–720.

Saad, Y. (2003). *Iterative methods for sparse linear systems*. Siam.

Sartori, I., & Hestnes, A. G. (2007). Energy use in the life cycle of conventional and low-energy buildings: A review article. *Energy and Buildings, 39*, 249–257.

Spatari, S., Betz, M., Florin, H., Baitz, M., & Faltenbacher, M. (2001). Using GaBi 3 to perform life cycle assessment and life cycle engineering. *The International Journal of Life Cycle Assessment, 6*, 81–84.

Spielmann, M., & Scholz, R. (2005). Life cycle inventories of transport services: Background data for freight transport (10 pp). *International Journal of Life Cycle Assessment, 10*, 85–94.

Suh, S. (2004). A note on the calculus for physical input–output analysis and its application to land appropriation of international trade activities. *Ecological Economics, 48*, 9–17.

Suh, S. (2005). Theory of materials and energy flow analysis in ecology and economics. *Ecological Modelling, 189*, 251–269.

Suh, S. (2009). *Handbook of input-output analysis economics in industrial ecology*. London: Springer.

Suh, S., & Heijungs, R. (2007). Power series expansion and structural analysis for life cycle assessment. *International Journal of Life Cycle Assessment, 12*, 381–390.

Suh, S., & Huppes, G. (2005). Methods for life cycle inventory of a product. *Journal of Cleaner Production, 13*, 687–697.

Suh, S., & Kagawa, S. (2005). Industrial ecology and input-output economics: An introduction. *Economic Systems Research, 17*, 349–364.

Suh, S., & Kagawa, S. (2009). Industrial ecology and input-output economics: A brief history. *Handbook of Input-Output Economics in Industrial Ecology*, 43–58 (Springer).

Suh, S., Lenzen, M., Treloar, G. J., Hondo, H., Horvath, A., Huppes, G., et al. (2004). System boundary selection in life-cycle inventories using hybrid approaches. *Environmental Science and Technology, 38*, 657–664.

Suh, S., & Nakamura, S. (2007). Five years in the area of input-output and hybrid LCA. *International Journal of Life Cycle Assessment, 12*, 351–352.

Szargut, J. (2005). Exergy method: Technical and ecological applications. WIT Press.

Szargut, J., Morris, D. R., & Steward, F. R. (1988). Exergy analysis of thermal, chemical, and metallurgical processes. Hemisphere.

Szargut, J., Ziębik, A., & Stanek, W. (2002). Depletion of the non-renewable natural exergy resources as a measure of the ecological cost. *Energy Conversion and Management, 43*, 1149–1163.

Tillman, A.-M., Ekvall, T., Baumann, H., & Rydberg, T. (1994). Choice of system boundaries in life cycle assessment. *Journal of Cleaner Production, 2*, 21–29.

Treloar, G. J. (1997). Extracting embodied energy paths from input–output tables: towards an input–output-based hybrid energy analysis method. *Economic Systems Research, 9*, 375–391.

Treloar, G. J. (1998). *Comprehensive embodied energy analysis framework*. Deakin University.

Treloar, G., Fay, R., Ilozor, B., & Love, P. (2001). An analysis of the embodied energy of office buildings by height. *Facilities, 19*, 204–214.

Tukker, A., Huppes, G., Oers, L., & Heijungs, R. (2006). *Environmentally extended input-output tables and models for Europe*.

UNI, E. (2001). 14040 (2006) *Environmental management—life cycle assessment—principles and framework*. International organisation for standardisation, July. Brown B, Aaron M.

United Nations. Statistical Division. (2008). *International Standard industrial classification of all economic activities (ISIC)* (Rev. 4th ed.). United Nations, New York.

Varga, R. S. (2009). *Matrix iterative analysis*. Springer Science & Business.

Watkins, D. S. (2004). *Fundamentals of matrix computations*. Hoboken: Wiley.

Waugh, F. V. (1950). Inversion of the Leontief matrix by power series. *Econometrica: Journal of the Econometric Society*, 142–154.

Weisz, H., & Duchin, F. (2006). Physical and monetary input–output analysis: What makes the difference? *Ecological Economics, 57*, 534–541.

Wiedmann, T., Minx, J., Barrett, J., & Wackernagel, M. (2006). Allocating ecological footprints to final consumption categories with input–output analysis. *Ecological Economics, 56*, 28–48.

Williams, E. (2004). Energy intensity of computer manufacturing: Hybrid assessment combining process and economic input-output methods. *Environmental Science and Technology, 38*, 6166–6174.

Wilting, H. C. (1996). *An energy perspective on economic activities*.

Wright, D. J. (1975). The natural resource requirements of commodities. *Applied Economics, 7*, 31–39.

Chapter 3
Accounting for Energy-Resources use by Thermodynamics

This chapter investigates the role played by thermodynamics in quantifying energy-resources use. Specifically, the main achievements of this chapter are: (1) to present a general overview of *thermodynamics-based life cycle methods*, and to propose a tentative taxonomy of such methods; (2) to define processes for the thermodynamic characterization of *non-renewable energy-resources* for the purpose of Input-Output analysis and environmental accountings in general.

What emerges from the current chapter is that many exergy based methods have been proposed in recent literature to account for natural resources requirements of productive processes and energy conversion systems. However, all these methods are not clearly and univocally defined from the methodological viewpoint. For this reason, exergy is adopted in the following chapters for the characterization of non-renewable energy-resources within the framework of Input-Output analysis.

3.1 Life Cycle Assessment methods based on Thermodynamics

Primary energy-resources are essential in order to sustain all the modern economic activities. Various kind of natural resources may be involved in production processes and transformed through them causing environmental impact: emissions to air, depletion of fossil fuels, raw materials consumption, land occupation, water use etc. Quantification of resources use is certainly important in order to evaluate the possible future scenarios in a sustainable perspective. However, to perform rigorous analyses, it is necessary to implement methods that comply with the basic scientific laws such as the *First* and the *Second Laws* of thermodynamics (Ukidwe et al. 2009). In this context, thermodynamic-based metrics, namely *energy*, *entropy generations* and *exergy*, allow to take into account a huge number of resources. This section aims at providing a general overview about the most commonly

© The Author(s) 2016
M.V. Rocco, *Primary Exergy Cost of Goods and Services*,
PoliMI SpringerBriefs, DOI 10.1007/978-3-319-43656-2_3

adopted thermodynamic-based methods that seek to quantify resources requirements over the life cycle of a generic system. Because of the huge literature produced in recent years about exergy based methods, special attention is devoted to such techniques and one tentative taxonomy of them is finally proposed.

3.1.1 Energy-Based methods

Early applications of thermodynamic concepts to life cycle assessment were proposed as two different methodologies contemporarily and independently developed: *Cumulative Energy Demand* and *Emergy analysis.*

The concept of *Cumulative Energy Demand* (CED) refers to the amount of primary energy required to deliver a good or service, considering both the direct and the indirect contributions along its production chain (Cleveland and Costanza 2007). In the early 1970s, these studies were the first concerning a LCA view and, usually, they account for non-renewable fossil fuels only. The objective of these analysis is to account for the *Energy Return on Investment* (EROI), defined as the ratio between the energy delivered by a process and the energy used directly and indirectly in that process (Cleveland and Costanza 2008). Early versions of Process analysis and Input-Output Analysis described in Chap. 2 were formulated and adopted in order to apply such method.

At about the same time, a very original line of thought was devised by *Odum* in its *Emergy analysis*: they adopted an embodied energy paradigm, but measuring all types of primary energy requirements by a conventional equivalent amount of solar radiation (Jorgensen et al. 2004; Odum 1994; Odum and Odum 2000). In contrast to other methods, Emergy has its own unit of measure: the *emjoule* (also called *solar emjoule*), to emphasize that primary solar energy contributions are taken into account. As for the exergy analysis, emergy analysis is able to account different and various forms of resources and it is able to consider the different quality of the energy flows (Hau and Bakshi 2004b) thanks to the concept of *transformity* (or *transformation ratio*), which is defined as *the solar emergy required to make one Joule of a service or product.* Detailed calculations of transformities for many different products are available in literature. Emergy is also able to take into account economic inputs and human resources, by introducing the ground-breaking idea that human work and monetary circulation in a society are in fact supported by the cumulative amount of emergy that the society could avail itself of. Based on this hypothesis, calculation of an *emergy/money ratio* was proposed (Ukidwe and Bakshi 2004). The Emergy analysis differs from traditional accounting approaches, establishing its own accounting rules (Brown and Herendeen 1996). As for all the accounting methods, Emergy analysis suffers from uncertainties, especially in the quantification of the transformities, and it presents allocation problems mostly related to co-products. In any case, Emergy analysis appears as the first attempt to unify the processes that occur in ecosystems and human activities (Hau and Bakshi 2004b).

3.1.2 Entropy-Based methods

The use of *entropy* in system analysis and optimization has a history that can be traced back to the early beginning of thermodynamics (Kondepudi and Prigogine 2014). The concept of entropy was introduced to describe the loss of work potential occurring in any energy conversion process, and only recently entropy analysis has been used to assess the overall resource consumption of one generic system. Two entropy based method emerges from literature: *Entropy Generation Minimization* (EGM) and *Cumulative Entropy production.*

The *Entropy Generation Minimization* (EMG) method, introduced by *Bejan* (1995, 2002), aims at evaluate the Second Law efficiency of a given energy conversion system, revealing the source of inefficiencies by identifying the large amount of entropy generations within the system and thus providing useful information for thermodynamic optimization. This method is designed to improve systems at a very detailed level: for such reason, it requires to know details about the geometrical setup of the system and the thermodynamic properties of the flows of matter and energy that cross its boundaries. However, since the boundaries defined by EGM are restricted to a specific process, its results exclude the sources of inefficiencies in supply chains: in order to fully assess the overall effects of optimizations at the process level, a different approach is then required (Bakshi et al. 2011).

Differently from EGM, the *Cumulative Entropy production* approach proposed by *Reisemann* consists in extending the boundaries of entropy analysis at the level of production and consumption systems that span several plants and regions. It is not so much aimed at optimizing the design of specific systems, but rather at assessing the resource consumption accompanying the economic production and consumption of goods in a Life Cycle perspective (Gößling-Reisemann 2006, 2008).

3.1.3 Exergy-Based methods

Exergy based methods for system analysis have been largely deepened in literature, as demonstrated by the huge number of scientific articles and books published every year. For such reason, the current paragraph devotes special attention to such methods.

Exergy is defined by literature as *the amount of useful work extractable from a generic system when it is brought to equilibrium with its reference environment through a series of reversible processes in which the system can only interact with such environment* (Kotas 2012; Moran and Sciubba 1994; Moran et al. 2010). Definition of reference environmental conditions is therefore an implicit prerequisite for the evaluation of exergy, which can be defined as a property of both the system and its surrounding environment. If the system is closed with respect to its reference environment, the final equilibrium state is called *Environmental State* (subscript 0), and implies a complete thermo-mechanical equilibrium ($T = T_0, p = p_0$). Once the environmental state has been reached, the chemical and/or phase composition of the system may be different from that of the environment and material exchanges may take place until the system reaches another thermodynamic state called *Dead State* (subscript 00), described by both thermo-mechanical and chemical equilibrium ($T = T_0, p = p_0, \mu_i = \mu_{i,0}$). The generalized exergy balance, analytically formulated among others by Bejan (2006), can be derived by combining energy and entropy balances and results as in relation (3.1).

$$\frac{dEx}{dt} = \sum_j \dot{Ex}_{W,j}^{\leftarrow} + \sum_k \dot{Ex}_{Q,j}^{\leftarrow} + \sum_q \dot{m}_q^{\leftarrow} ex_q - \dot{Ex}_D \qquad (3.1)$$

According to the classification proposed by *Kotas*, definitions of non-flow exergy (3.2), exergy related to work (3.3), heat (3.4) and bulk flow (3.5) interactions are introduced. Exergy of bulk flow (3.5) can be expressed as the sum of four components: potential (*pt*), kinetic (*kn*), physical (*ph*) and chemical (*ch*) exergy.

$$Ex = E - T_0 S + p_0 V \qquad (3.2)$$

$$\dot{Ex}_W^{\leftarrow} = \dot{W}^{\leftarrow} \qquad (3.3)$$

$$\dot{Ex}_Q^{\leftarrow} = \dot{Q}^{\leftarrow}(1 - T_0/T_k) \qquad (3.4)$$

$$\begin{aligned} ex &= ex_{pt} + ex_{kn} + ex_{ph} + ex_{ch} \\ &= g\Delta z + \frac{1}{2}\Delta w^2 + \Delta h - T_0 \Delta s + \sum_i \Delta g_i y_i \end{aligned} \qquad (3.5)$$

Finally, the exergy destruction term $\dot{Ex}_D$ represents the irreversible loss of work capacity experienced by the system during a thermodynamic process, and it is evaluated through the so-called *Gouy-Stodola theorem* (Kotas 2012), as in relation (3.6).

$$\dot{Ex}_D = \dot{W} - \dot{W}_{rev} = T_0 \dot{S}_{gen} \qquad (3.6)$$

Because of its features, exergy is widely recognized by literature as a convenient tool for system analysis and optimization, and as a suited indicator for quantification of resource consumption (Kotas 1985; Moran and Sciubba 1994).

In recent years, many different *embodied exergy* methods have been proposed by the literature in order to extend temporal and spatial boundaries of conventional exergy analysis to account for requirements of primary resources (Stougie and Van der Kooi 2011). All the embodied exergy methods proposed by the literature have their roots in standard exergy analysis and they may be classified in the broad category collectively known as *Thermoeconomics*, defined as the discipline that merges *thermodynamics* and *cost accounting* principles (Rocco et al. 2014a). The word *Thermoeconomics* was coined in 1961 by *Tribus* (Evans and Tribus 1965), and further fundamental developments were made by *El-Sayed* and *Evans* (1970) in US and by *Elsner, Fratzscher, Beyer* and *Brodjansky* in Europe (Elsner and Fratzscher 1957). Later, in the 80s, *Gaggioli* extended the application of the theory to encompass a broader set of energy-intensive systems (Gaggioli 1983). Recently *Valero* et al. (Lozano and Valero 1993; Valero 1989; Valero et al. 1986a, b, c), were able to produce a complete and elegant formalization of the method, that is now known as *Exergy Cost theory*. For a review of the developments of the theory and of its applications, see (Bejan et al. 1996; Erlach et al. 1999; Sciubba and Wall 2010; Valero et al. 2010).

In 1986, *Szargut* and *Morris* propose the *Cumulative Exergy Consumption* (CExC), also known as the *Cumulative Exergy Demand* (CExD) to account for the overall primary resources use in the production of a good or a service (Szargut 2005a; Szargut et al. 1988). CExC method is similar to Cumulative Energy Consumption, but it relies on exergy instead of energy. The application of CExC requires to expand the boundaries of the production system by encompassing all the industrial processes needed to convert both renewable and non-renewable natural resources into the desired product (Hau and Bakshi 2004a). The use of exergy allows to take into account not only energy flows that cross system boundaries, but also other types of primary resources such as water, metals and minerals (Bösch et al. 2007; Szargut and Stanek 2007; Szargut et al. 2002). Based on CExC concept, *Szargut* and *Stanek* later introduced the *Thermo-Ecological Cost* (TEC) defined as the cumulative consumption of non-renewable primary exergy required for the production phase of the considered good or service (Szargut and Stanek 2007).

A life cycle extension of the above introduced methods was proposed by *Cornelissen* in 1997 through the *Exergy Life Cycle Assessment* (ELCA) (Cornelissen 1997; Cornelissen and Hirs 2002; RL et al.). This approach extends temporal and spatial boundaries of CExC analysis, evaluating materials and energy requirements for all the life cycle phases of the considered system by means of their exergy equivalents. A further extension of the ELCA, called *Zero-ELCA*, was proposed by *Cornelissen* in order to include the impact due to pollutant emissions into the primary exergy requirements of the considered product, by including the additional primary exergy requirements of emission abatement processes into the accounting (Hirs 2003; RL et al.; Zhou et al. 2013).

Recently, *Dewulf* proposes the *Cumulative Exergy Extraction from Natural Environment* (CEENE) to account for the primary renewable and non-renewable resources invoked for the production of a generic good or service, including water, minerals and metals extraction as for CExC. Differently from CExC, CEENE also accounts for the primary exergy requirements due to land occupation: indeed, from a purely ecological perspective, when land is occupied the ecosystem is deprived of the solar exergy necessary to sustain its natural cycles. Therefore, cumulative land use is taken into account as a solar exergy contribution (Dewulf et al. 2007).

Other developments of CExC method were proposed by *Ukiwide* and *Bakshi* (Hau and Bakshi 2004a) named *Industrial/Ecological Cumulative Exergy*

Consumption (ICEC/ECEC). ICEC considers only the non-renewable natural resources requirements (expressed by means of exergy) of the analyzed productive process, and it is then very close to the TEC method. A complementary result is then obtained through the ECEC method, which adds to results of ICEC also the exergy consumed by ecological processes to produce primary flows of energy, materials and to dissipate the emissions. Notice that, differently from *all* the other method described in this section, ICEC/ECEC method is explicitly formalized by means of the same mathematics of Input-Output analysis (Ukidwe et al. 2009). ECEC method is conceptually close to the Emergy method, trying to expand the boundary over the ecosystem processes. Many practical applications of ICEC/ECEC method can be found in literature (Hau and Bakshi 2004a; Ukidwe and Bakshi 2007; Zhang et al. 2010a, b). As for the Emergy analysis, ECEC method lacks in the allocation of embodied exergy to multiple products, and large uncertainties are caused by the lack of data for most of the processes that occur within ecological systems. Moreover, the formulation of this method do not encompasses all the life cycle of the system: exergy requirements due to disposal of the product and the influence of recycling are not taken into account (Ukidwe et al. 2009).

The *Extended Exergy Accounting* (EEA) was conceived by *Sciubba* in 1998 (Sciubba 2001), and it can be considered as a further extension of the ELCA method: the objective of EEA consists in the evaluation of primary exergy requirements of the whole life cycle of a given product (Sciubba 2005), including the side-effects caused by externalities such as human labor, capital circulation and environmental pollution. Extensive review of EEA, with methodological details and practical applications can be found in literature (Dai et al. 2014; Hang et al. 2016; Rocco et al. 2014a; Sciubba 2011, 2013; Sciubba and Zullo 2011; Talens Peiró et al. 2010). According to the literature, and with reference to relation (3.7), the Extended Exergy of the generic ith product can be evaluated for each of its life cycle phases as the sum of its Cumulative Exergy Consumption plus the additional contributions of primary exergy requirements due to externalities, including human labor EE_L, capitals circulation EE_K and environmental remediation EE_O.

$$EE_i = \int\limits_{t=LC_i} (CExC_i + EE_{ext})_j dt \rightarrow EE_{ext} = EE_L + EE_K + EE_O \qquad (3.7)$$

The environmental remediation costs EE_O are evaluated as the additional Extended Exergy expenditures required by real pollutants treatment processes, in a similar manner as introduced in the Zero-LCA approach. On the other hand, evaluation of labour and capital externalities requires the introduction of the following two fundamental postulates:

1. *In any society, a portion of the gross global influx of primary exergy Ex_{in} is used to sustain the workers who generate labour.* According to this postulate, the exergy equivalent of one working hour ee_L can be estimated as the ratio between the exergy converted into labor EE_L and the total number of hours cumulatively produced by the society in the same time frame, as in relation (3.8).

$$EE_L = ee_L N_{wh} \quad \rightarrow \quad ee_L = \left(\frac{HDI}{HDI_o}\right) ex_{surv} \frac{N_h}{N_{wh}} \tag{3.8}$$

According to relation (3.8), *HDI* and *HDI*$_0$ are respectively the Human Development Index of the considered country and of the reference pre-industrial society (about 0.055); e_{surv} represents the minimum exergy amount required for the metabolic survival of an individual (about 1.05×10^7J/p · *day*) and N_h/N_{wh} is the ratio between the cumulative amount of living hours and working hours of the considered country;

2. *The amount of exergy required to generate the monetary circulation within a society is proportional to the amount of exergy embodied into labour.* This postulate establishes a link between the working hours production and the monetary circulation within a specific country: this allows to account for the exergy embodied in one monetary unit through relation (3.9).

$$EE_K = EE_L \cdot \left(\frac{M_f}{S}\right) \quad \rightarrow \quad ee_K = EE_L \cdot \left(\frac{M_f}{S}\right) \cdot \frac{1}{M2} \tag{3.9}$$

According to relation (3.9), the exergy embodied in monetary circulation EE_K is related to EE_L through the ratio between the monetary circulation due to financial activities M_f and the gross cumulative wages S. The exergy embodied in one monetary unit is then evaluated dividing EE_K by the total monetary circulation within the considered country $M2$.

The outcome of the literature analysis above presented results in the tentative taxonomy of embodied exergy methods, proposed in Table 3.1.

Classification of the methods is performed according to three main categories:

1. *Primary factors* category identifies all the natural resources that are taken into account by each method. Classification of primary resources is provided by LCA guidelines (Curran 2012; Guinée 2002; Pennington et al. 2004);
2. *Secondary factors* category identifies the products for which the respective embodied exergy is taken into account by the analysis method;
3. *Life Cycle phases* category identifies the phases accounted for by the analysis. Notice that the sub-category named *Production of NR resources* refers to the extension of spatial and temporal boundaries of the method in order to account for the primary resources required by the environment to produce the non-renewable resources through its natural cycles.

Detailed theoretical treatment and applications of all the exergy based methods is out of the scope of the book and can be found in literature (Bakshi et al. 2011; Orsi et al. 2016; Rocco et al. 2014b, 2016; Stougie and Van der Kooi 2011). The fundamental element that emerges from literature is that research efforts related to thermodynamic system analysis seems to be generally focused on the development of LCA methods based on exergy for the quantification of primary natural resources consumption due to goods and services production. Finally, literature highlights

Table 3.1 A proposed taxonomy for embodied exergy methods

Category	Type	CExC	TEC	CEENE	ELCA	Z-ELCA	ICEC	ECEC	EEA
Primary factors	Fossil fuels (NR)	●	●	●	●	●	●		●
	Nuclear (NR)	●	●	●	●	●	●		●
	Kinetic	●		●	●	●			●
	Solar	●		●	●	●		●	●
	Potential	●		●	●	●			●
	Biomass (NR)	●	●	●	●	●	●		●
	Biomass	●		●	●	●			●
	Water	●		●	●	●			●
	Metals (NR)	●		●	●	●		●	●
	Minerals (NR)	●		●	●	●		●	●
	Land use			●					
Secondary factors	Energy carriers	●	●	●	●	●	●	●	●
	Goods and services	●	●	●	●	●	●	●	●
	Monetary capitals						●	●	●
	Working hours								●
Life Cycle phases	Construction	●	●	●	●	●	●	●	●
	Operation			●	●	●			●
	Disposal			●	●	●			●
	Production of NR res.							●	
	Environmental remed.					●			●

that even if the temporal and spatial domains and the primary factors included in the evaluation are clearly defined for all the analyzed methods, resources accounting techniques are not properly and univocally established, making applications of such methods very difficult and not properly standardized.

3.2 Thermodynamic characterization of non-renewable energy-resources

In order to quantify the amount of primary non-renewable energy-resources embodied in goods and services through the Input-Output analysis presented in Chap. 2, a proper thermodynamic characterization of such resources should be performed. In principle, energy-resources could be simply measured by means of their *mass* or *volume* (e.g. *tons* of raw coal, Nm^3 of natural gas, etc.). However, different energy-resources are usually characterized by different physical and thermodynamic properties, and thus the sum of their mass or volume could produce misleading results. To overcome this issue, literature suggests to characterize energy-resources by means of the thermodynamic effects they generate through a defined reference depletion process: (1) *Heat* or *Entropy generation* caused by a *spontaneous* (i.e. irreversible) depletion process; (2) *Mechanical work* produced through a *reversible* depletion process (Gyftopoulos and Beretta 2005). All these quantities can be derived by manipulating standard *energy* and *entropy* balances (3.10) and (3.11).

$$\frac{dE}{dt} = \sum_j \dot{W}_j^\leftarrow + \dot{Q}_0^\leftarrow + \sum_k \dot{Q}_k^\leftarrow + \sum_q \dot{m}_q^\leftarrow \left(gz + \frac{w^2}{2} + h \right)_q \tag{3.10}$$

$$\frac{dS}{dt} = \frac{\dot{Q}_0^\leftarrow}{T_0} + \sum_k \frac{\dot{Q}_k^\leftarrow}{T_k} + \sum_q \dot{m}_q^\leftarrow s_q + \dot{S}_{gen} \tag{3.11}$$

As will be showed in the following, apart for the definition of one specific *depletion process* (reversible or irreversible), these characterization methods *always* requires definition of one *reference environment*: usually, *Standard State* is assumed for such purpose as a temperature T_0 of 298.15 K (25 °C), pressure P_0 of 100.0 kPa and a chemical composition of reference air, oceans and earth crust defined by *Szargut* et al. (2005).

3.2.1   Irreversible depletion process

Let us consider the combustion chamber of Fig. 3.1, such that:

1. The system operates at steady state;
2. Flows of *fuel* (*f*), *stoichiometric dry air* (*a*) and *flue gases* (*fg*) are entering/exiting the combustion chamber at environmental temperature T_0 and pressure p_0;
3. The inlet dry air (*a*) provides theoretical amount of oxygen for the complete combustion of the fuel;
4. Apart from the bulk flow interactions, the system exchange heat $\dot{Q}_0^\leftarrow$ with the environment, crossing an ideal boundary at temperature T_0.

According to these hypotheses, application of the energy balance (3.10) allows to evaluate the amount of heat produced by combustion of a mole or mass unit of fuel delivered to the environment $q_0^\rightarrow$, expressed by relation (3.12): literature refers to such quantity as the *Heating Value* of the fuel. Notice that the heat is delivered to the environment through an ideal surface at temperature T_0.

$$\dot{Q}_0^\rightarrow = (\dot{H}_R - \dot{H}_P)_{T_0, p_0} \rightarrow q_0^\rightarrow = HV = \frac{\dot{Q}_0^\rightarrow}{\dot{n}_f} \tag{3.12}$$

Fig. 3.1 Steady state spontaneous process used for the evaluation of heating value and entropy generation of fossil fuels

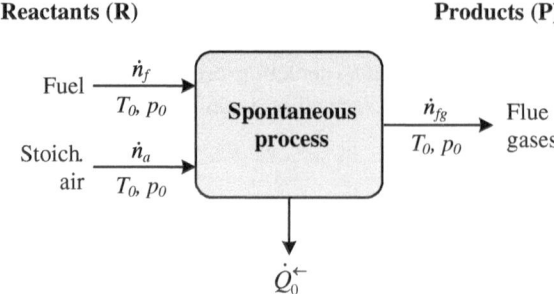

Since the process is assumed as *spontaneous* it is also *irreversible*, and application of combined energy and entropy balance lead to the evaluation of the maximum amount of entropy that could be generated by depleting the mole or mass unit of fuel, as described by relation (3.13).

$$\begin{cases} \dot{Q}_0^{\rightarrow} = (\dot{H}_R - \dot{H}_P)_{T_0,p_0} \\ \frac{\dot{Q}_0^{\leftarrow}}{T_0} + (\dot{S}_R - \dot{S}_P)_{T_0,p_0} + \dot{S}_{gen} = 0 \end{cases} \rightarrow \dot{S}_{gen} = \frac{\dot{Q}_0^{\leftarrow}}{T_0} - (\dot{S}_R - \dot{S}_P) \rightarrow s_{gen} = \frac{\dot{S}_{gen}^{\rightarrow}}{\dot{n}_f}$$

$$(3.13)$$

3.2.2 Reversible depletion process

The combustion chamber of Fig. 3.2 is now considered and analyzed according to the same hypotheses introduced in Sect. 3.2.1. However, in this case the fuel is depleted through a reversible process: the entropy generation term in balance (3.11) is set to zero, the system exchange heat $\dot{Q}_0^{\rightarrow}$ with the environment through an ideal surface at temperature T_0 and it produces also mechanical work $\dot{W}_{rev}^{\rightarrow}$.

Fig. 3.2 Steady state reversible process used for the evaluation of availability (or exergy) of fossil fuels

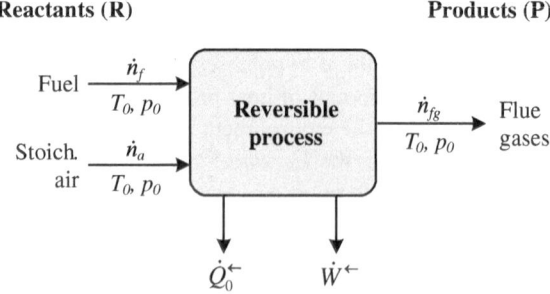

Because the outlet flow stream is in equilibrium with the reference environment with respect to temperature and pressure, $w_{rev}^{\rightarrow}$ represent the *availability*, or *exergy*, of the mole or mass unit of the considered fuel and it is expressed by relation (3.14).

$$\begin{cases} \dot{Q}_0^{\leftarrow} + \dot{W}^{\leftarrow} + \left(\dot{H}_R - \dot{H}_P\right)_{T_0,p_0} = 0 \\ \frac{\dot{Q}_0^{\leftarrow}}{T_0} + \left(\dot{S}_R - \dot{S}_P\right)_{T_0,p_0} = 0 \end{cases} \rightarrow \dot{W}_{rev}^{\rightarrow} = \dot{Ex}^{\rightarrow} = \dot{H}_R - \dot{H}_P - T_0 \cdot \left(\dot{S}_R - \dot{S}_P\right) \tag{3.14}$$

Before calculating *heating value HV*, *entropy generation* s_{gen} or *chemical exergy* ex_{ch} of the considered fuel, it is required to define specific models and assumptions for the calculation of enthalpy and entropy of material flows entering and exiting the ideal reactors of Figs. 3.1, 3.2.

With respect to energy, exergy allows to homogenously account for both the quantity and quality of material and energy interactions, referring them to a reversible thermodynamic exploitation process. Moreover, exergy seems to be characterized by lower sensibility to environmental conditions with respect to entropy generations. For these reasons, exergy has been accepted by scientific community as the most suited metric to account for natural resources consumption, and it has been assumed as the quantitative basis for Industrial Ecology (Bakshi et al. 2011; Seager and Theis 2002). Recently, also the *International Reference Life Cycle Data System* (ILCD) refers to exergy as *the best and most mature midpoint indicator to account for resources consumption* (Michael et al. 2011). However, it has to be remarked that the interpretation of exergy as a measure of the thermodynamic value of a substance makes sense only for substances that are somehow used to drive a process. Indeed, the absolute exergy value of substances has relevance only when they are used energetically (which includes, e.g., driving a chemical reaction). In case of substances that are used non-energetically (e.g., by providing structures or by conducting heat and electricity), their exergy content has no practical relevance. However, for substances or materials that are used in a cascading way, the relevance of their exergy content might, however, come into play in a later-use phase (Gößling-Reisemann 2006, 2008; Seager and Theis 2002).

3.2.3 The Exergy of common fossil fuels

Detailed analytical methods for the calculation of chemical exergy of organic and inorganic substances were developed by *Szargut* et al. (Morris and Szargut 1986; Szargut 1989, 2005b; Szargut et al. 1987), *Kotas* (2013), *Gyftopoulos* and *Beretta* (2005), *Moran* and *Shapiro* (Moran et al. 2010). Thermo-physical properties of *gaseous fuels*, either pure or as mixtures, can be easily derived through the ideal or real gas models: for this reason, evaluation of standard chemical exergy of gaseous fuels can be performed starting from the analytical expression (3.14). This is not the

case of complex *liquid* and *solid fuels* (e.g. coal, oil or biomass), for which literature
has developed specific correlations. Among others, *Song* et al. propose the unified
correlation (3.15) for estimating specific standard chemical exergy of solid and
liquid fuels (Song et al. 2012).

$$ex_{ch}(T_0, p_0) = 363.439C + 1075.633H - 86.308O + 4.147N + 190.798S - 21.1A$$
$$(3.15)$$

Given the weight content, in percentage, of carbon $(27.33\% \leq C \leq 89.10\%)$,
hydrogen $(2.46\% \leq H \leq 14.00\%)$, oxygen $(1.10\% \leq O \leq 46.92\%)$, nitrogen
$(0.00\% \leq N \leq 9.27\%)$, sulphur $(0.01\% \leq S \leq 5.54\%)$ and ash $(0.00\% \leq A \leq 51.96\%)$
of the fuel, correlation (3.15) returns its chemical exergy expressed in kJ/kg. Such
correlation is applicable to coal, biomass, petroleum, shale oil, oil from tar sands,
crude benzol, synthetic liquid fuels made from coal or biomass. Average heating
value and chemical exergy of the common primary non-renewable fossil fuels are
reported in Table 3.2. Notice that the so-called *Szargut factor* β defined by relation
(3.16) is the ratio between Chemical Exergy and Heating Values of the fuel.

$$\beta = \frac{ex_{ch}}{HV} \qquad (3.16)$$

In the following chapters, values collected in Table 3.2 will be used as the
reference for primary fossil fuels characterization.

Table 3.2 Heating value and exergy of the common non-renewable fuels. Biomass category can
be classified as renewable or non-renewable

Impact category	Resource	Units	HV	ex_{ch}	β
Fossil	Coal, brown, in ground	MJ/kg	9.9	10.3	1.04
	Coal, hard, unspecified, in ground	MJ/kg	19.1	19.7	1.03
	Gas, mine, off-gas, process, coal mining	MJ/m³	39.8	37.4	0.94
	Gas, mine, off-gas, process, coal mining	MJ/kg	49.8	46.8	0.94
	Gas, natural, in ground	MJ/m³	38.3	36	0.94
	Oil, crude, in ground	MJ/kg	45.8	46.5	1.02
	Peat, in ground	MJ/kg	9.8	10.3	1.05
Nuclear	Uranium, in ground	MJ/kg	560,000	560,000	1.00
Biomass	Wood, hard, standing	MJ/m³	12,740	13,377	1.05
	Wood, soft, standing	MJ/m³	9180	9639	1.05
	Wood, unspecified, standing	MJ/m³	10,960	11,508	1.05

3.2.4 The role of reference environment

One of the most debated issues related to thermodynamic characterization of energy-resources concerns the influence that *environmental reference conditions* have on the calculation of heating value, entropy generation or chemical exergy of fossil fuels. Indeed, the real environment is far from equilibrium and involves in changings of its properties—especially its temperature—that are functions of seasons and geographical location.

As can be inferred from Sect. 3.2.3 thermodynamic characterization of fossil fuels requires the definition of one reference depletion process, and this imply the definition of one reference environment in turn. Indeed, exploitation of one resource can be seen as the extraction of thermodynamic utility (heat or work) through an ideal process that brought the resource in physical and chemical equilibrium with the environment. The assumption that the environment is in equilibrium at constant temperature, pressure and chemical composition is thus necessary in order quantify heating value, entropy generation or chemical exergy of fuels. Once a reference environment has been defined (here, the *Standard State:* $T_{ref} = 25\,°C$, $p_{ref} = 1\,bar$, denoted by the superscript °), the question about the influence that a change in environmental temperature T_0 has on such numerical values arise. As an example, this paragraph evaluate how such values are sensible to a change in environmental temperature between -10 and $50\,°C$ for 1 kmol of pure methane (CH_4) that is involved in the stoichiometric combustion reaction (3.17).

$$CH_4 + 2(O_2 + 3.76N_2) \quad \rightarrow \quad CO_2 + 2H_2O + 7.52N_2 \qquad (3.17)$$

Molar values of *enthalpy* and *absolute entropy* for any given pure substance at the environmental temperature and pressure can be evaluated according to the following relations (Moran et al. 2010).

$$
\begin{aligned}
\bar{h}(T_0, p_0) &= \bar{h}_f^° + \left[\bar{h}(T_0, p_0) - \bar{h}(T_{ref}, p_{ref}) \right] \\
\bar{s}(T_0, p_0) &= \bar{s}^° + \left[\bar{s}(T_0, p_0) - \bar{s}(T_{ref}, p_{ref}) \right]
\end{aligned}
\qquad (3.18)
$$

Assuming *ideal gas* behavior, the above introduced relations can be rewritten as (3.19), where $R = 8.314\,kJ/kmolK$ is the universal gas constant and $\bar{c}_p$ is the molar isobaric heat capacity, evaluated according to the *NASA* polynomial expressions (Moran et al. 2010) and $\bar{h}_f^°$ is the molar enthalpy of formation.

$$
\begin{aligned}
\bar{h}(T_0, p_0) &= \bar{h}_f^° + \bar{c}_p(T_0 - T_{ref}) \\
\bar{s}(T_0, p_0) &= \bar{s}^° + \bar{c}_p \, \ln\frac{T_0}{T_{ref}} - R \, \ln\frac{p_0}{p_{ref}}
\end{aligned}
\qquad (3.19)
$$

Enthalpy and entropy of an ideal mixture of ideal gases results as in relation (3.20), where y_i denotes the molar composition of the mixture.

$$\bar{h}_{mix}(T_0, p_0) = \sum_i y_i \cdot \bar{h}_i(T_0, p_0)$$

$$\bar{s}_{mix}(T_0, p_0) = \sum_i y_i \cdot \bar{s}_i(T_0, p_0) - R \cdot \sum_i y_i \cdot \ln y_i \tag{3.20}$$

Heating value (3.21), *entropy generations* (3.22) and *chemical exergy* (3.23) of 1 kmol of pure methane can be thus evaluated as functions of the environmental temperature T_0.

$$HV(T_0, p_0) = \Delta h_r^o + \sum_R \left[v_i \bar{c}_{p,i}(T_0 - T_{ref}) \right]_R - \sum_P \left[v_j \bar{c}_{p,j}(T_0 - T_{ref}) \right]_P \tag{3.21}$$

$$\bar{s}_{gen}(T_0, p_0) = \frac{HV(T_0, p_0)}{T_0} - \left[\sum_i v_i \cdot \bar{s}_i(T_0, p_0) - R \cdot \sum_i v_i \cdot \ln y_i \right]_{R-P} \tag{3.22}$$

$$\overline{ex}_{ch}(T_0, p_0) = HV(T_0, p_0) - T_0 \left[\sum_i v_i \cdot \bar{s}_i(T_0, p_0) - R \cdot \sum_i v_i \cdot \ln y_i \right]_{R-P} \tag{3.23}$$

Where P and R respectively refers to *products* and *reactants,* and v are the stoichiometric coefficients of the combustion reaction (3.17).

For practical reasons, fossil fuels are usually characterized with respect to the *Standard State*. However, according to these relations, it can be said that heating value, entropy generation and chemical exergy of ideal gas mixtures of fossil fuels are all sensible to changes in the environmental temperature, as showed by Table 3.3. Considering the case of methane, entropy generation results in a stronger dependence with respect to heating value or chemical exergy. Relative differences (Δ, in percentage) between results obtained at T_0 and at T_{ref} are depicted in Fig. 3.3. Heating Value and Chemical exergy are not sensible to environmental temperature variation, whereas Entropy generation undergoes in relevant changes.

Table 3.3 Sensitivity of heating value, entropy generation and chemical exergy of methane to the environmental temperature

T_0 (°C)	HV (MJ/kg)	ΔHV (%)	s_{gen} (MJ/kgK)	Δs_{gen} (%)	ex_{ch} (MJ/kg)	Δex_{ch} (%)
−10	50.03	0.04	196.52	12.81	51.71	−0.43
−5	50.03	0.04	192.97	10.77	51.75	−0.37
5	50.02	0.03	186.27	6.92	51.81	−0.25
15	50.02	0.01	180.03	3.34	51.87	−0.12
25 *(ref)*	*50.01*	–	*174.20*	–	*51.94*	–
40	50.00	−0.02	166.17	−4.61	52.04	0.19
50	50.00	−0.03	161.23	−7.45	52.10	0.31

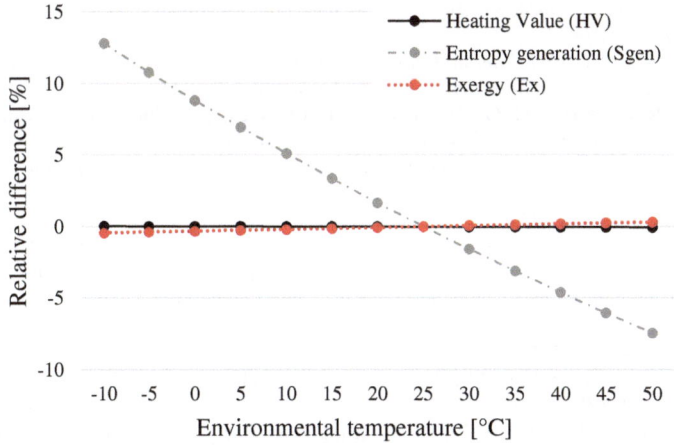

Fig. 3.3 Sensitivity of heating value, entropy generation and chemical exergy of methane to the environmental temperature

Intuitively, using IOA for the evaluation of primary energy-resources embodied in goods and services requires to compile the exogenous transactions matrix by means of heating value, entropy generation or chemical exergy of fossil fuels. However, due to the assumption of *process aggregation* described in Sect. 2.3.4, it may happen that productive processes operating in different geographical locations (with different environmental temperatures) are ideally aggregated into one single process: therefore, using entropy generation to characterize gaseous fossil fuels may produce errors in estimating the embodied resources in products. However, further investigations are required before extending such theoretical results to all the fossil fuels.

References

Bakshi, B. R., Gutowski, T. G. P., & Sekulić, D. P. (2011). *Thermodynamics and the destruction of resources*. New York: Cambridge University Press.

Bejan, A. (1995). *Entropy generation minimization: The method of thermodynamic optimization of finite-size systems and finite-time processes*. CRC press.

Bejan, A. (2002). Fundamentals of exergy analysis, entropy generation minimization, and the generation of flow architecture. *International Journal of Energy Research, 26*, 0–43.

Bejan, A. (2006). *Advanced engineering thermodynamics*. Wiley.

Bejan, A., Tsatsaronis, G., Moran, M. J. (1996). *Thermal design and optimization*. Wiley.

Bösch, M. E., Hellweg, S., Huijbregts, M. A., & Frischknecht, R. (2007). Applying cumulative exergy demand (CExD) indicators to the ecoinvent database. *International Journal of Life Cycle Assessment, 12*, 181–190.

Brown, M. T., & Herendeen, R. A. (1996). Embodied energy analysis and EMERGY analysis: A comparative view. *Ecological Economics, 19*, 219–235.

Cleveland, C., & Costanza, R. (2007). Net energy analysis. In *Encyclopedia of the Earth*.

Cleveland, C. J., & Costanza, R. (2008). Energy return on investment (EROI). In *Encyclopedia of Earth (online), April.*

Cornelissen, R. L., (1997). *Thermodynamics and sustainable development; the use of exergy analysis and the reduction of irreversibility.* Universiteit Twente.

Cornelissen, R. L., & Hirs, G. G. (2002). The value of the exergetic life cycle assessment besides the LCA. *Energy Conversion and Management, 43,* 1417–1424.

Curran, M. A. (2012). *Life cycle assessment handbook: A guide for environmentally sustainable products.*

Dai, J., Chen, B., & Sciubba, E. (2014). Ecological accounting based on extended exergy: A sustainability perspective. *Environmental Science and Technology, 48,* 9826–9833.

Dewulf, J., Bösch, M., Meester, B. D., Vorst, G. V. D., Langenhove, H. V., Hellweg, S., et al. (2007). Cumulative exergy extraction from the natural environment (CEENE): A comprehensive life cycle impact assessment method for resource accounting. *Environmental Science and Technology, 41,* 8477–8483.

El-Sayed, Y., & Evans, R. B. (1970). Thermoeconomics and the design of heat systems. *Journal of Engineering for Power, 92,* 27.

Elsner, N., & Fratzscher, W. (1957). Die Bedeutung der Exergieflußbilder für die Untersuchung wärmetechnischer Anlagen, gezeigt am Beispiel eines Abhitzekessels, eines Wärmekraftwerks und einer Dampflokomotive.

Erlach, B., Serra, L., & Valero, A. (1999). Structural theory as standard for thermoeconomics. *Energy Conversion and Management, 40,* 1627–1649.

Evans, R. B., & Tribus, M. (1965). Thermo-economics of saline water conversion. *Industrial & Engineering Chemistry Process Design and Development, 4,* 195–206.

Gaggioli, R. A. (1983). *Efficiency and costing: Second law analysis of processes.* American Chemical Society.

Gößling-Reisemann, S. (2006). Entropy as a measure for resource consumption—application to primary and secondary copper production. In *Sustainable metals management.* Springer, pp. 195–235.

Gößling-Reisemann, S. (2008). What is resource consumption and how can it be measured? *Journal of Industrial Ecology, 12,* 10–25.

Guinée, J. B. (2002). Handbook on life cycle assessment operational guide to the ISO standards. *The International Journal of Life Cycle Assessment, 7,* 311–313.

Gyftopoulos, E. P. & Beretta, G. P. (2005). *Thermodynamics: Foundations and applications.* Courier Corporation.

Hang, M. Y. L. P., Martinez-Hernandez, E., Leach, M., & Yang, A. (2016). Towards a coherent multi-level framework for resource accounting. *Journal of Cleaner Production, 125,* 204–215.

Hau, J. L., & Bakshi, B. R. (2004a). Expanding exergy analysis to account for ecosystem products and services. *Environmental Science and Technology, 38,* 3768–3777.

Hau, J. L., & Bakshi, B. R. (2004b). Promise and problems of emergy analysis. *Ecological Modelling, 178,* 215–225.

Hirs, G. (2003). Thermodynamics applied. Where? Why? *Energy, 28,* 1303–1313.

Jorgensen, S., Odum, H., & Brown, M. (2004). Emergy and exergy stored in genetic information. *Ecological Modelling, 178,* 11–16.

Kondepudi, D., & Prigogine, I. (2014). *Modern thermodynamics: From heat engines to dissipative structures.* Wiley.

Kotas, T. J. (1985). *The exergy method of thermal plant analysis.* Butterworths.

Kotas, T. J. (2012). *The exergy method of thermal plant analysis.* Paragon Publishing.

Kotas, T. J. (2013). *The exergy method of thermal plant analysis.* Elsevier.

Lozano, M., & Valero, A. (1993). Theory of the exergetic cost. *Energy, 18,* 939–960.

Michael, H., Mark, G., Jerome, G., Reinout, H., Mark, H., Olivier, J., Manuele, M., & An, D. S. (2011). *Recommendations for life cycle impact assessment in the European context—based on existing environmental impact assessment models and factors (International Reference Life Cycle Data System—ILCD handbook).* Publications Office of the European Union.

Moran, M., & Sciubba, E. (1994). Exergy analysis: Principles and practice. *ASME Transactions Journal of Engineering Gas Turbines and Power, 116*, 285–290.

Moran, M. J., Shapiro, H. N., Boettner, D. D., & Bailey, M. (2010). *Fundamentals of engineering thermodynamics*. Wiley.

Morris, D. R., & Szargut, J. (1986). Standard chemical exergy of some elements and compounds on the planet earth. *Energy, 11*, 733–755.

Odum, H. T. (1994). The emergy of natural capital. In *Investing in natural capital: The ecological economics approach to sustainability* (pp. 200–214). Washington (DC): Island Press.

Odum, H. T., & Odum, E. P. (2000). The energetic basis for valuation of ecosystem services. *Ecosystems, 3*, 21–23.

Orsi, F., Muratori, M., Rocco, M., Colombo, E., & Rizzoni, G. (2016). A multi-dimensional well-to-wheels analysis of passenger vehicles in different regions: Primary energy consumption, CO_2 emissions, and economic cost. *Applied Energy, 169*, 197–209.

Pennington, D., Potting, J., Finnveden, G., Lindeijer, E., Jolliet, O., Rydberg, T., et al. (2004). Life cycle assessment part 2: Current impact assessment practice. *Environment International, 30*, 721–739.

RL, C., van Nimwegen, A., & Hirs,Cornelissen R.L., Van Nimwegen P.A., Hirs G.G. (2000). Exergetic life-cycle analysis. *In Proceedings of ECOS 2000, Enschende, Netherlands*.

Rocco, M., Colombo, E., & Sciubba, E. (2014a). Advances in exergy analysis: A novel assessment of the extended exergy accounting method. *Applied Energy, 113*, 1405–1420.

Rocco, M. V., Cassetti, G., Gardumi, F., & Colomb, E. (2016). Exergy life cycle assessment of soil erosion remediation technologies: An Italian case study. *Journal of Cleaner Production, 112*, 3007–3017.

Rocco, M. V., Colombo, E., & Sciubba, E. (2014b). Advances in exergy analysis: A novel assessment of the extended exergy accounting method. *Applied Energy, 113*, 1405–1420.

Sciubba, E. (2001). Beyond thermoeconomics? The concept of extended exergy accounting and its application to the analysis and design of thermal systems. *Exergy, an International Journal, 1*, 68–84.

Sciubba, E. (2005). Exergo-economics: Thermodynamic foundation for a more rational resource use. *International Journal of Energy Research, 29*, 613–636.

Sciubba, E. (2011). A revised calculation of the econometric factors α- and β for the extended exergy accounting method. *Ecological Modelling, 222*, 1060–1066.

Sciubba, E. (2013). Can an environmental indicator valid both at the local and global scales be derived on a thermodynamic basis? *Ecological Indicators, 29*, 125–137.

Sciubba, E., & Wall, G. (2010). A brief commented history of exergy from the beginnings to 2004. *International Journal of Thermodynamics, 10*, 1–26.

Sciubba, E., & Zullo, F. (2011). Is sustainability a thermodynamic concept? *International Journal of Exergy, 8*, 68–85.

Seager, T., & Theis, T. (2002). A uniform definition and quantitative basis for industrial ecology. *Journal of Cleaner Production, 10*, 225–235.

Song, G., Xiao, J., Zhao, H., & Shen, L. (2012). A unified correlation for estimating specific chemical exergy of solid and liquid fuels. *Energy, 40*, 164–173.

Stougie, L., & Van der Kooi, H. (2011). The relation between exergy and sustainability according to literature. In C. J. Koroneos, D. C. Rovas & A. T. Dompros (Eds.), ELCAS2011: Proceedings of the 2nd International Exergy, Life Cycle Assessment and Sustainability Workshop and Symposium, 19–21 June 2011, Nisyros Island, Greece, pp. 590–597.

Szargut, J. (1989). Chemical exergies of the elements. *Applied Energy, 32*, 269–286.

Szargut, J. (2005a). *Exergy method: Technical and ecological applications*. WIT Press.

Szargut, J. (2005b). *Exergy method: Technical and ecological applications*. WIT press.

Szargut, J., Morris, D. R., & Steward, F. R. (1987). *Exergy analysis of thermal, chemical, and metallurgical processes*.

Szargut, J., Morris, D. R., & Steward, F. R. (1988). *Exergy analysis of thermal, chemical, and metallurgical processes*. Hemisphere.

Szargut, J., & Stanek, W. (2007). Thermo-ecological optimization of a solar collector. *Energy, 32,* 584–590.

Szargut, J., Valero, A., Stanek, W., & Valero, A. (2005). Towards an international legal reference environment. *Proceedings of ECOS, 2005,* 409–420.

Szargut, J., Ziębik, A., & Stanek, W. (2002). Depletion of the non-renewable natural exergy resources as a measure of the ecological cost. *Energy Conversion and Management, 43,* 1149–1163.

Talens Peiró, L., Villalba Méndez, G., Sciubba, E., & i Durany, X. G. (2010). Extended exergy accounting applied to biodiesel production. *Energy, 35,* 2861–2869.

Tsatsaronis, G. (1993). Thermoeconomic analysis and optimization of energy systems. *Progress in Energy and Combustion Science, 19,* 227–257.

Ukidwe, N. U., & Bakshi, B. R. (2004). Thermodynamic accounting of ecosystem contribution to economic sectors with application to 1992 US economy. *Environmental Science and Technology, 38,* 4810–4827.

Ukidwe, N. U., & Bakshi, B. R. (2007). Industrial and ecological cumulative exergy consumption of the United States via the 1997 input–output benchmark model. *Energy, 32,* 1560–1592.

Ukidwe, N. U., Hau, J. L., & Bakshi, B. R. (2009). Thermodynamic input-output analysis of economic and ecological systems. *Handbook of input-output economics in industrial ecology* (pp. 459–490). Springer.

Valero, A. (1989). Thermoeconomics: A new chapter of physics. In *Workshop on Energy Analysis of Power Plants, Pisa Auditorium ENEL-CRTN, Italy,* Feb.

Valero, A., Lozano, M., & Muñoz, M. (1986a). A general theory of exergy saving. I. On the exergetic cost. *Computer-Aided Engineering and Energy Systems: Second Law Analysis and Modelling, 3,* 1–8.

Valero, A., Lozano, M., & Muñoz, M. (1986b). A general theory of exergy saving. II. On the thermoeconomic cost. *Computer-Aided Engineering and Energy Systems: Second Law Analysis and Modelling, 3,* 1–8.

Valero, A., Lozano, M., & Muñoz, M. (1986c). A general theory of exergy saving. III. Energy saving and Thermoeconomics. *Computer-Aided Engineering and Energy Systems: Second Law Analysis and Modelling, 3,* 1–8.

Valero, A., Usón, S., Torres, C., & Valero, A. (2010). Application of thermoeconomics to industrial ecology. *Entropy, 12,* 591–612.

Zhang, Y., Baral, A., & Bakshi, B. R. (2010a). Accounting for ecosystem services in life cycle assessment, part II: Toward an ecologically based LCA. *Environmental Science and Technology, 44,* 2624–2631.

Zhang, Y., Singh, S., & Bakshi, B. R. (2010b). Accounting for ecosystem services in life cycle assessment, part I: A critical review. *Environmental Science and Technology, 44,* 2232–2242.

Zhou, R., Liu, C., Li, J., & Yu, J. X. (2013). ELCA evaluation for keyword search on probabilistic XML data. *World Wide Web, 16,* 171–193.

Chapter 4
Exergy based Input-Output analysis

In this chapter, the *Exergy based Input-Output analysis* (ExIO) is defined by the joint application of Input-Output analysis and Exergy analysis. ExIO may be helpful in solving drawbacks highlighted by the literature (see Sect. 1.2). The use of exergy within the mathematical framework of Input-Output Analysis can be performed in two ways:

1. *Characterization of exogenous transactions matrix*. In this case, exogenous transactions matrix is defined by means of exergy, whereas the system IOT is defined in monetary, physical or hybrid units. Application of Leontief model to such system results in the evaluation of the embodied exergy (i.e. the exergy cost) of goods and services produced by a generic production system in a more meaningful way with respect to other metrics. Because the heating value and the exergy of fossil fuels result very close to each other (Gyftopoulos and Beretta 1991; Kotas 2012), values of embodied energy and exergy will result numerically similar;

2. *Characterization of both exogenous transactions matrix and input-output table*. Exergy is adopted to characterize all the environmentally extended Input-Output table. This practice allows to perform Exergy Cost analysis of the considered system, identifying and quantifying the exergy destructions and their relative impacts on the definition of embodied exergy of products, thus giving a better thermodynamic insight on the considered system.

The *Exergy based Input-Output analysis* (ExIO) is here defined as a tool for the evaluation of primary non-renewable exergy embodied in goods and services of a given national economy, relying on standard and constantly updated data sources (Sect. 4.1). Hybrid Input-Output models are then introduced in order to increase the accuracy of results, enabling to evaluate the embodied exergy of detailed products within a certain economy (Sect. 4.2). The evaluation of thermodynamics performances of energy conversion systems based on Exergy Cost analysis and the Exergy Life Cycle Assessment is finally discussed (Sect. 4.3).

© The Author(s) 2016   61
M.V. Rocco, *Primary Exergy Cost of Goods and Services*,
PoliMI SpringerBriefs, DOI 10.1007/978-3-319-43656-2_4

4.1 Evaluating primary exergy embodied in goods and services

As discussed in Sect. 2.3.2, the application of Leontief model to a generic productive system results in specific and total demand of exogenous transactions invoked for the production of each final demand product. However, in modern economies, exogenous resources are rarely taken directly from the environment: most likely, they are produced outside system boundaries by other systems. For this reason, the evaluation of *primary* resources requirements through IOA requires in principle to extend the boundaries of system IOT till all the exogenous transactions are directly taken from the environment, as shown in Fig. 4.1.

Compilation of input-output tables starting from scratch, encompassing all the supply chains of the considered system, could be a very difficult and time intensive way to evaluate primary resources embodied in products. Moreover, due to the arbitrariness in the definition of supply chains boundaries and due to uncertainties in definition of appropriate cutoff criteria, this practice makes results of different analyses of a same product not comparable. It follows that rules for definition of the system boundaries and a convenient and standardized representation of the supply chains need to be established. In order to reach such goal, literature suggests to rely on *Monetary Input—Output Tables* (MIOTs) of *national economies* (Hendrickson et al. 1997, 2010; Joshi 1999). Indeed, MIOTs provide a *standard, largely available* and *constantly updated* classification of all the economic and productive activities within a given national economy. Given a generic national MIOT, if the exogenous transactions matrix is defined as the primary fossil fuels entering the economy, calculation of the exergy embodied in national products can be performed by applying the Leontief model as shown in Sect. 2.3. Evaluating environmental burdens of production activities through MIOTs is a common practice known in literature as the *Economic Input-Output LCA* (EIO-LCA) (Matthews and Small

Fig. 4.1 Extension of system IOT to account for primary exogenous resources embodied in the generic target product

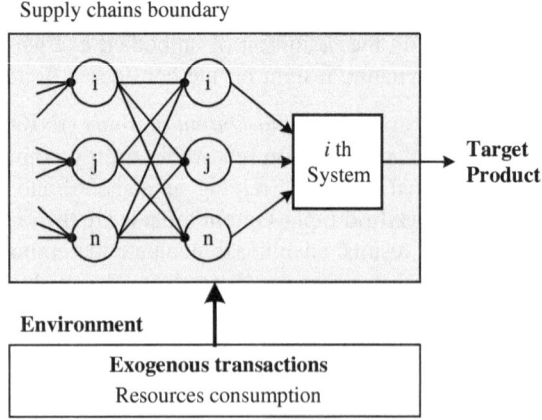

2000). The general structure of a MIOT and its use in the evaluation of embodied exergy of products is presented in Sect. 4.1.1.

Although the use of MIOTs is recognized by the literature as a fundamental step forward in LCA, its application presents two major flaws: (1) national economies are actually not closed clusters of productive processes, thus specific assumptions are required to treat *international trades of goods and services;* (2) the national economic activities are usually collected into broad sectors with *high level of aggregation.* Inappropriate treatment of such issues may produce wrong or inaccurate results.

4.1.1 Monetary Input-Output tables of national economies

Input-Output analysis was originally conceived with the fundamental purpose to analyze the interdependence of economic sectors within a certain economic system (Dietzenbacher and Lahr 2004; Leontief 1986). Nowadays, developments of the IOA framework initially set by Leontief in late 1930s are key components of many economic and environmental analyses (Baumol 2000; Miller and Blair 2009b).

General structure of a MIOT

The MIOT of one given nation has the same structure of the IOT described in paragraph 2.3.2: all the economic and productive activities are subdivided into distinct *productive sectors*, or *segments*, that produce goods or services for intermediate consumption and for final demand (Miller and Blair 2009b). MIOTs may be compiled according to different degree of aggregation, depending on data availability and on the purpose of IOA: productive sectors may be represented as whole industrial activities (e.g. "manufacturing sector"), specific production industries (e.g. "steel production") or even much smaller categories (e.g. "production of steel nails and spikes"). As its name suggests, the MIOT is compiled exclusively in monetary values. With respect to physical or hybrid units, describing products trades within a national economy using monetary values is considered as the most rational choice for many practical reasons:

- Products of national economies are, except for few cases like *Developing Countries* (DCs), composed by a very large number of both material and immaterial products as output of each productive sector. It is therefore very difficult to find a unique physical unit suited to account for each sector, especially in case of high level of aggregation;
- Products trades within a country are periodically collected by means of their monetary equivalents rather than in physical units: compiling an IOT in physical or hybrid units could be then a very challenging task;
- MIOTs are constantly updated and compiled according to international standards, such as the *International Standard Industrial Classification of all economic activities* (ISIC) (United Nations. Statistical Division 2008) or the *Statistical Classification of Economic Activities in the European Country*

(NACE) (Eurostat 2008). Such tables are mainly used for different kind of economic analyses and environmental accounts (also defined by literature as the *satellite accounts*) (Tukker et al. 2006).

The structure of a MIOT of a generic national economy in a defined time frame (usually one year) is presented in Fig. 4.2: the economy is ideally divided into distinct sectors, represented by one row and one column. All the direct relations among each sector constitutes the *national transaction matrix*, in which every line represents the products sales of one sector to the other sectors and every column represents the products that one sector purchases from the others.

With reference to Fig. 4.2, in addition to the endogenous transactions matrix $\mathbf{Z}$, final demand matrix $\mathbf{f}$ and total production vector $\mathbf{x}$ (shaded squares), the MIOT is also composed by the *value added matrix* $\mathbf{v}$ and the *imports matrix* $\mathbf{m}$. Components of the final demand matrix, given by expression (4.1), are respectively *household purchases* (h_i), *purchases for (private) investment purposes* (i_i), *government purchases* (g_i) (federal, public, and local), and *exports* (ex_i), each of one referred to the ith productive sector.

$$\mathbf{f}(n \times 4) = \begin{bmatrix} \mathbf{f_H} & \mathbf{f_I} & \mathbf{f_G} & \mathbf{f_E} \end{bmatrix} \rightarrow \mathbf{f} = \begin{bmatrix} h_i & i_i & g_i & ex_i \\ \vdots & \vdots & \vdots & \vdots \\ h_n & i_n & g_n & ex_n \end{bmatrix} \qquad (4.1)$$

Value added matrix $\mathbf{v}$, given by expression (4.2), is composed by payments for labor compensation (l_i) and government services (taxes), interests on invested capitals, land, entrepreneurship profit and other minor voices (n_i). Value added can be generated by the producing sectors expenses ($\mathbf{v_Z}$) or directly by household and govern outlays ($\mathbf{v_f}$).

$$\mathbf{v} = \begin{bmatrix} \mathbf{v_Z} | \mathbf{v_f} \end{bmatrix} = \begin{bmatrix} l_1 & \cdots & l_n & l_H & l_I & l_G & l_E \\ n_1 & \cdots & n_n & n_H & n_I & n_G & n_E \end{bmatrix} \qquad (4.2)$$

In a similar way, total imports matrix $\mathbf{m}$ (4.3) can be expressed as the sum of imported products invoked for intermediate requirements ($\mathbf{m_Z}$) and for final demand ($\mathbf{m_f}$).

$$\mathbf{m} = \begin{bmatrix} \mathbf{m_Z} | \mathbf{m_f} \end{bmatrix} = \begin{bmatrix} m_1 & \cdots & m_n & m_H & m_I & m_G & m_E \end{bmatrix} \qquad (4.3)$$

For each ith productive sectors of an n sectors economy, it is possible to write the economic production balance (4.4): it accounts for the production of the ith good or service for intermediate and final uses.

$$x_i = \sum_{j=1}^{n} x_{ij} + f_i \quad with: \quad f_i = h_i + i_i + g_i + ex_i \qquad (4.4)$$

	1	...	n	Final demand				Total output
Sector 1	x	...	x	h_1	i_1	g_1	x_1	X_1
...	⋮	⋱	⋮	⋮	⋮	⋮	⋮	⋮
Sector n	x	...	x_{nn}	h_n	i_n	g_n	x_n	x_n
Value added	l_1	...	l_n	l_H	l_I	l_G	l_E	L
	n_1	...	n_n	n_H	n_I	n_G	n_E	N
Imports	m_1	...	m_n	m_H	m_I	m_G	m_E	M
Total outlays	X_1	...	X_n	H	I	G	E	

Fig. 4.2 MIOT of a generic national economy (Miller and Blair 2009b)

Considering the endogenous production only (without consider goods or services imports), *total output* of every sector can be evaluated through expression (4.4): this expression can be rewritten in matrix form as showed by relation (4.5), where vector **x** represents the economic value produced by each sector in the considered time window.

$$\mathbf{x} = \mathbf{Z}\mathbf{i} + \mathbf{f}\,\mathbf{i}$$

$$
\begin{bmatrix} x_1 \\ \vdots \\ x_n \end{bmatrix} =
\begin{bmatrix} x_{11} & \cdots & x_{1n} \\ \vdots & \ddots & \vdots \\ x_{n1} & \cdots & x_{nn} \end{bmatrix} \cdot \mathbf{i}\,(n \times 1) +
\begin{bmatrix} h_i & i_i & g_i & ex_i \\ \vdots & \vdots & \vdots & \vdots \\ h_n & i_n & g_n & ex_n \end{bmatrix} \cdot \mathbf{i}\,(4 \times 1)
$$

$$(4.5)$$

Conversely, the *total outlays* of the *i*th sector are defined as the sum of products value required from other sectors, value added and imports, and they result as the row sum of the corresponding matrices, expressed by relation (4.6).

$$\mathbf{x}^T = \mathbf{Z}^T\mathbf{i} + \mathbf{v}_Z^T\,\mathbf{i} + \mathbf{m}_Z^T$$

$$
\begin{bmatrix} x_1 \\ \vdots \\ x_n \end{bmatrix} =
\begin{bmatrix} x_{11} & \cdots & x_{n1} \\ \vdots & \ddots & \vdots \\ x_{1n} & \cdots & x_{nn} \end{bmatrix} \cdot \mathbf{i}(n \times 1) +
\begin{bmatrix} l_1 & n_1 \\ \vdots & \vdots \\ l_n & n_n \end{bmatrix} \cdot \mathbf{i}(2 \times 1) +
\begin{bmatrix} m_1 \\ \vdots \\ m_n \end{bmatrix}
\quad (4.6)
$$

In MIOTs, the equality between total output and total outlays holds: for every *i*th sector, the economic output produced equals the sum of purchases, value added and imports. This equality could be extended to all the economy introducing total value added $L+N$ and total imports M (as column sum), and the total domestic final demand $H + I + G$ and total exports E (as row sum). In this perspective, relation (4.7) expresses the *circular* nature of the economy (Miller and Blair 2009b), in which the total value of output equals the total value of outlays and the economy is said to be balanced.

$$\mathbf{i}(n \times 1) \cdot \mathbf{x} + L + N + M = \mathbf{i}(n \times 1) \cdot \mathbf{x} + H + I + G + E = X \qquad (4.7)$$

If boundaries of MIOTs are defined as the border of the national economies, the total amount of *endogenous* final demand (including exports and the portion of value added directly generated by household and govern outlays) is the monetary aggregate known as *Gross Domestic Product* (GDP), as in relation (4.8). On the other hand, the *Gross National Product* (GNI) results if the MIOT includes all the products produced by all the enterprises owned by the citizens of the country (no matter where they actually are). In other words, GDP defines its scope according to location, while GNI defines its scope according to ownership (Kendrick 1996).

$$GP = \mathbf{i}(1 \times n) \cdot \left\{ \begin{bmatrix} \mathbf{f} \\ \mathbf{v_f} \end{bmatrix} [(n+2) \times 1] \cdot \mathbf{i}(4 \times 1) \right\} \quad \rightarrow \quad GP = \begin{cases} GDP \\ GNP \end{cases} \qquad (4.8)$$

A large number of nations routinely publish their MIOTs according to internationally agreed standard set of recommendations of the *System of National Accounts* (SNA), which describes macroeconomic accounts in the context of internationally agreed concepts, definitions and rules (Nations et al. 2009). Further details about the possible uses of MIOTs for economic and statistics purposes are out of the scope of this book and are they can be found in literature (Kendrick 1996; Miller and Blair 2009b). MIOTs provide available, constantly updated and standardized data sources for the application of both Leontief model. For these reasons, use of MIOTs is recognized by the literature as the most convenient, reliable and simple way for the evaluation of primary resources embodied in products (Ferrão and Nhambiu 2009; Hendrickson et al. 2010).

A focus on international trades of goods and services
One of the main limitations in the application of Leontief model to MIOTs resides in the treatment of international trades of goods and services. Indeed, national economies are actually not closed clusters because of such trades of products, as shown by Fig. 4.3:

• *Imported products*: exergy embodied in the whole national final demand does not consist only in the primary exergy directly extracted by the considered economy, but it also includes further contributions related to extraction, processing and transporting of imported products in the economy (Bullard III et al. 1975; Nakamura et al. 2007). Such imported products come from unknown processes, and specific assumptions are required to estimate their contributions in terms of embodied exergy;

- *Exported products.* Although MIOTs usually consider exported products as part of the final demand, such trades could be actually part of both intermediate consumption and final demand of other countries.

With reference to Fig. 4.2, MIOTs are usually compiled as the assembly of two main tables: the endogenous table (subscript *end*) and the imports table (subscript *imp*). The national endogenous Input-Output table is given by the assembly of transaction matrix $\mathbf{Z}_{end}$, final demand vector $\mathbf{f}_{end}$ and total production vector $\mathbf{x}_{end}$, for which the endogenous production balance (4.9) holds.

$$\mathbf{x}_{end} = \mathbf{Z}_{end} \cdot \mathbf{i} + \mathbf{f}_{end} \cdot \mathbf{i} \qquad (4.9)$$

On the other hand, national Input-Output table for the imports side is given by the assembly of intermediate imports matrix $\mathbf{Z}_{imp}$, final demand imports vector $\mathbf{f}_{imp}$ and total imports vector $\mathbf{x}_{imp}$. Generally, it is convenient to distinguish among imports that are also endogenously produced, or *competitive imports* (subscript *c*), and those that are not domestically produced, or *noncompetitive imports* (subscript *nc*). Relation (4.3) can be rewritten and expanded according to these definitions as in relation (4.10). Usually, it is assumed that there are no imported products (both competitive and non-competitive) that are also exported, that is, the last column in relation (4.10) is empty.

$$\mathbf{m} = \begin{bmatrix} \mathbf{m}_{c,Z} & \mathbf{m}_{c,f} \\ \mathbf{m}_{nc,Z} & \mathbf{m}_{nc,f} \end{bmatrix} = \begin{bmatrix} m_{c,11} & \cdots & m_{c,1n} & m_{c,Hi} & m_{c,Ii} & m_{c,Gi} & m_{c,Ei} \\ \vdots & \ddots & \vdots & \vdots & \vdots & \vdots & \vdots \\ m_{c,n1} & \cdots & m_{c,nn} & m_{c,Hn} & m_{c,In} & m_{c,Gn} & m_{c,En} \\ m_{nc,1} & \cdots & m_{nc,n} & m_{nc,H} & m_{nc,I} & m_{nc,G} & m_{nc,E} \end{bmatrix}$$

$$(4.10)$$

National imported production balance can be written as in relation (4.11).

$$\mathbf{x}_{imp} = \mathbf{Z}_{imp} \cdot \mathbf{i} + \mathbf{f}_{imp} \cdot \mathbf{i}$$

$$\begin{bmatrix} \mathbf{x}^{c}_{imp}(n \times 1) \\ \mathbf{x}^{nc}_{imp}(1 \times 1) \end{bmatrix} = \begin{bmatrix} \mathbf{Z}^{c}_{imp}(n \times n) \\ \mathbf{Z}^{nc}_{imp}(1 \times n) \end{bmatrix} \cdot \mathbf{i}(n \times 1) + \begin{bmatrix} \mathbf{f}^{c}_{imp}(n \times 4) \\ \mathbf{f}^{nc}_{imp}(1 \times 4) \end{bmatrix} \cdot \mathbf{i}(4 \times 1) \qquad (4.11)$$

Noncompetitive imports are usually internalized in one category of *competitive* imports: this assumption is realistic for large economies, for which all imported products are also endogenously produced, and for which one standard classification

Fig. 4.3 Representation of a generic national economy through the MIOT

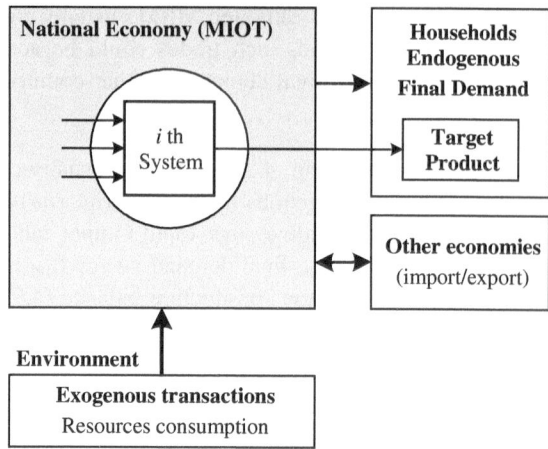

of all the economic activities is adopted. In this case, import matrix (4.10) can be simplified as in relation (4.12).

$$
\mathbf{m} = [\mathbf{m_Z} | \mathbf{m_f}] = \begin{bmatrix} m_{c,11} & \cdots & m_{c,1n} & m_{c,Hi} & m_{c,Ii} & m_{c,Gi} & m_{c,Ei} \\ \vdots & \ddots & \vdots & \vdots & \vdots & \vdots & \vdots \\ m_{c,n1} & \cdots & m_{c,nn} & m_{c,Hn} & m_{c,In} & m_{c,Gn} & m_{c,En} \end{bmatrix}
$$

$$(4.12)$$

The MIOT of a national economy can be thus described by relation (4.13), while the total production balance for the generic national economy N can be defined in compact form as in relation (4.14).

$$
\begin{bmatrix} \mathbf{x_{end}} \\ \mathbf{x_{imp}} \end{bmatrix} = \begin{bmatrix} \mathbf{Z_{end}} \\ \mathbf{Z_{imp}} \end{bmatrix} \cdot \mathbf{i}(n \times 1) + \begin{bmatrix} \mathbf{f_{end}} \\ \mathbf{f_{imp}} \end{bmatrix} \cdot \mathbf{i}(4 \times 1) \tag{4.13}
$$

$$
[\mathbf{Z_N} | \mathbf{f_N}] = [\mathbf{Z_{end}} | \mathbf{f_{end}}] + [\mathbf{Z_{imp}} | \mathbf{f_{imp}}] \rightarrow \mathbf{x_N} = \mathbf{Z_N} \cdot \mathbf{i}(n \times 1) + \mathbf{f_N} \tag{4.14}
$$

$$
\mathbf{f_N} = (\mathbf{f_{end}} + \mathbf{f_{imp}}) \cdot \mathbf{i}(4 \times 1) \tag{4.15}
$$

In order to simplify the mathematical formulation of Input-Output analysis, the national final demand $\mathbf{f_N}$ will be expressed in the following by the column sum vector of relation (4.15).

Essential structure of a MIOT for the application of Exergy based Input-Output analysis

Based on the previously introduced concepts, to account for the primary exergy embodied in products of national economies it is required to define the *Environmentally extended Input-Output table* as depicted in Fig. 4.4: grey and white matrices represents respectively the endogenous and imported transactions within the considered economy. Accounting for embodied non-renewable exergy in goods and services produced within one given national economy requires the application of Leontief model through relations (2.18) and (2.28) (see Sect. 2.3.2) to the assembly given by the national MIOT and the national exogenous transactions matrix $\mathbf{R_{end}}$ (green matrix in Fig. 4.4). Specifically, $\mathbf{R_{end}}$ collects the amount of primary energy-resources produced within the national economy (i.e. endogenous extraction), such as raw coal, crude oil and natural gas expressed by means of their exergy equivalents. Notice that the ISIC/NACE classification defines the *Mining and Quarrying* sector as the interface between the national economy and the environment with respect to the primary non-renewable resources, that is, the sector in which primary non-renewable fossil fuels enter the economy (United Nations. Statistical Division 2008).

Due to the increasing amount of international trades of products, definition of appropriate ways to account for the exergy embodied in trades is nowadays crucial. Several international trades treatment models emerge from the literature and can be grouped into two main approaches: *Single-Region* and *Multi-Regional* models (Turner et al. 2007; Wiedmann 2009; Wiedmann et al. 2007). Practical ways to apply such models are formalized in the following paragraphs, and are schematically depicted in Fig. 4.5. Other possible approaches to apply such models can be found in literature: however, such customized methods can be always expressed as modified versions of the general models presented below (Ahmad and Wyckoff 2003; Lenzen et al. 2004).

4.1.2 Treatment of international trades: Single-Region models

The simplest way to account for the exergy embodied in the final demand of a given national economy N is by considering imports as if they were produced with the same efficiency of endogenous products (i.e. endogenously produced), while exports are entirely considered as a part of the final demand. These assumptions are also known in the literature as the *mirrored economy* (Strømman and Gauteplass 2004), or *autonomous region* (Lenzen et al. 2004). These models are justified by the fact that imported products would require this amount of primary exergy if they were manufactured within the considered economy (see Fig. 4.5, left side) (Bullard et al. 1978; Bullard III et al. 1975; Joshi 1999, 2000). Alternative methodological definitions of the methods formalized below can be retrieved in literature (Treloar 1997, 1998).

Model a—Endogenous economy
This is the easiest way to account for embodied exergy of national products: it consists in the application of Leontief model to the endogenous national economy

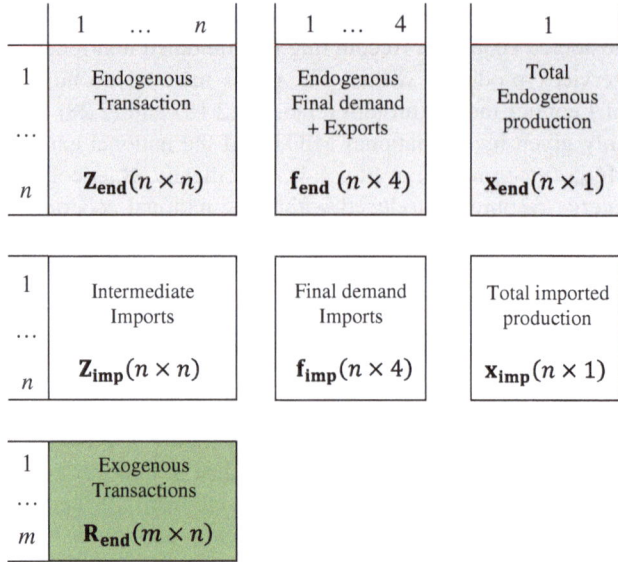

Fig. 4.4 Essential structure of a MIOT for the purpose of exergy based input-output analysis

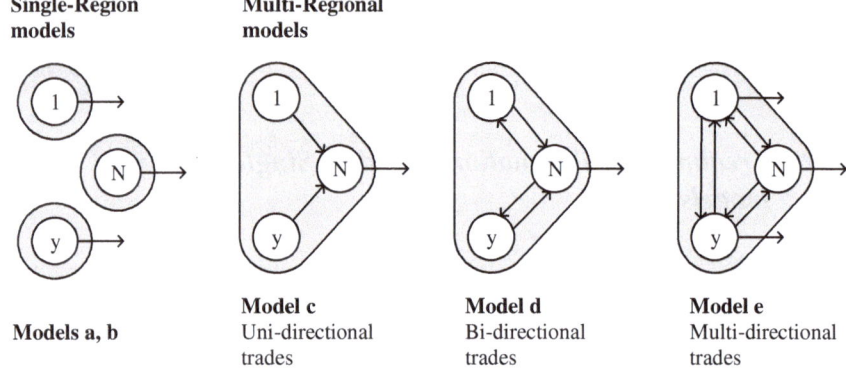

Fig. 4.5 Schematic representations of single-region and multi-regional models for the generic cluster composed by N national economies. Adapted from (Lenzen et al. 2004)

only, that is, to the national endogenous IOT represented by Fig. 4.4. The input coefficients matrix $\mathbf{B_a}$ is evaluated as the amount of the primary fossil fuels endogenously produced within the national economy $\mathbf{R_{end}}$, quantified by means of exergy, that are required to produce domestic goods and services $\mathbf{x_{end}}$. In a similar fashion, technical coefficients matrix $\mathbf{A_a}$ is evaluated considering only the endogenous economy, as shown by relation (4.16).

$$\mathbf{B_a} = \mathbf{R_{end}}\hat{\mathbf{x}}_{end}^{-1}; \quad \mathbf{A_a} = \mathbf{Z_{end}}\hat{\mathbf{x}}_{end}^{-1} \tag{4.16}$$

The evaluation of embodied exergy results from the application of Leontief model, as in relation (4.17). Notice that elements of $\mathbf{e_a}$ represent values of exergy embodied in the monetary unit of endogenous products, while elements of $\mathbf{E_a}$ represent the total amount of exergy embodied in national final demand $\mathbf{f_N}$ (including both endogenous and imported products).

$$\mathbf{L_a} = (\mathbf{I} - \mathbf{A_a})^{-1} \rightarrow \begin{cases} \mathbf{e_a} = (\mathbf{B_a}\mathbf{L_a})^{\mathrm{T}} \\ \mathbf{E_a} = \hat{\mathbf{f}}_N\,\mathbf{e_a} \end{cases} \tag{4.17}$$

Notice that *model a* is always derived in literature starting from the total national production balance defined by relation (4.14), and embodied environmental burdens are derived as the difference between total and imported production (Battjes et al. 1998; Miller and Blair 2009a).

Model b—Endogenous economy with constant exogenous input coefficients
This model has been introduced an alternative to *model a*. Differently from *model a*, technical coefficients matrix $\mathbf{A_b}$ is evaluated considering the total national production balance, including imported products. On the other hand, the input matrix $\mathbf{B_b}$ is evaluated as the amount of the primary fossil fuels endogenously produced $\mathbf{R_{end}}$ required to deliver domestic goods and services $\mathbf{x_{end}}$, as for *model a*. This practice is justified by the fact that the *efficiency* of production activities is defined as the amount of exogenous resources required to deliver one unit of product, thus it is actually given by input coefficients b_{kj}. With respect to *model a*, considering both domestic and imported products in the evaluation of Technical coefficients matrix should result in a better representation of the real productive structure of the considered economy (Turner et al. 2007).

$$\mathbf{B_b} = \mathbf{R_{end}}\hat{\mathbf{x}}_{end}^{-1}; \quad \mathbf{A_b} = \mathbf{Z_N}\hat{\mathbf{x}}_N^{-1} \tag{4.18}$$

Calculation of specific and total embodied exergy of national products results by relation (4.19).

$$\mathbf{L_b} = (\mathbf{I} - \mathbf{A_b})^{-1} \rightarrow \begin{cases} \mathbf{e_b} = (\mathbf{B_b}\mathbf{L_b})^{\mathrm{T}} \\ \mathbf{E_b} = \hat{\mathbf{f}}_N\,\mathbf{e_b} \end{cases} \tag{4.19}$$

4.1.3 Treatment of international trades: Multi-regional models

From the theoretical standpoint, an exact treatment of international trades of products requires to define the input-output table as closed with respect to imports and exports. This could be done by extending the boundaries of the considered national system till the products that cross these boundaries only consist in primary exergy directly extracted from natural environment (i.e. fossil fuels endogenously produced). This can be practically performed aggregating endogenous monetary input-output tables of different national economies, and by tracing the origin of imports and the destination of exports for each national economy. For this purpose, all the considered national MIOTs must be defined with the same aggregation level of the economic activities, and data should be expressed according to the same currency and properly harmonized. These models are usually defined as the *Multi-Regional IO models* (MRIO) originally described by Isard (1951, 1960), Isard and Langford (1971), Leontief (1953), Chereny (1953), Moses (1955) and others. MRIO tables of World regions were originally defined to analyze scenarios about future economic development and to evaluate possible alternative scenarios by means of emissions, energy use and mineral extraction (Duchin and Levine 2013; Leontief 1974). Many applications of MRIO models can be found in literature (Lenzen et al. 2004; Wiedmann et al. 2011).

MRIO models can be defined in two general ways:

- *Open MRIO models*. If the model does not include all the World economic activities, it is defined as *open* with respect to the other exogenous trades. Once such model has been defined, one of the Single-Region models previously introduced in Sect. 4.1.2 should be adopted to treat the remaining trades. Notably, the larger is the region covered by the MRIO model, the less relevant is the role played by such remaining exogenous trades (Battjes et al. 1998; Bullard III et al. 1975);

- *Closed MRIO models or World models*. If the model encloses all the World economies, it is also closed to all other exogenous trades of products and it is said to be a *closed* MRIO model (Duchin et al. 1996).

The international trades treatment models here described can be applied for the analysis of both the open and closed MRIO models. However, for the sake of simplicity, only closed models will be considered. In the following, subscripts ij refer to the transactions of products from ith to jth national economy (i.e. elements with subscript ii account for endogenous transactions).

Model c—Multi-regional model with uni-directional trades

Let us consider the entire World economy as an aggregation of a number of R national economies, each one represented by a MIOT as in Fig. 4.2, characterized by a number n of economic activities, compiled according to the same standards, assumptions and currency. The total number of producing sectors of the global economy is defined as $n_{tot} = n \cdot R$.

The uni-directional trades model, here *model c*, is proposed by the literature as the simplest way to account for resources/wastes embodied in products (Lenzen et al. 2004; Nijdam et al. 2005; Peters and Hertwich 2005, 2006; Round 2001; Weber and Matthews 2008). According to this model, the analyzed jth national economy receives trades from other economies for intermediate requirements $\mathbf{Z_{ij}}$

and for final demand $\mathbf{f_{ij}} = (\mathbf{ex_{ij}} - \mathbf{Z_{ij}i})$, while for the other $R - 1$ economies the remaining imported/exported products are internalized in the domestic production in a similar way as for *model b*. The World transaction matrix $\mathbf{Z_c}(n_{tot} \times n_{tot})$, final demand matrix $\mathbf{f_c}(n_{tot} \times R)$ and total production vector $\mathbf{x_c}(n_{tot} \times 1)$ are defined, and the *model c* can be described by the production balance (4.20).

$$\mathbf{x_c} = \mathbf{Z_c i_Z}(n_{tot} \times 1) + \mathbf{f_c i_f}(R \times 1)$$

$$
\begin{bmatrix} x_1 \\ x_j \\ x_R \end{bmatrix} = \begin{bmatrix} Z_{11} & Z_{1j} & - \\ - & Z_{jj} & - \\ - & Z_{Rj} & Z_{RR} \end{bmatrix} i_Z + \begin{bmatrix} f_{11} & (ex_{1j} - Z_{1j}i) & - \\ - & f_{jj} & - \\ - & (ex_{Rj} - Z_{Rj}i) & f_{RR} \end{bmatrix} i_f \quad (4.20)
$$

The application of Leontief model to the World production balance (4.20) requires he World exogenous transactions matrix $\mathbf{R_c}(m \times n_{tot})$ to be defined as the bordered matrix (4.21), composed by the exogenous resources matrix of each country $\mathbf{R_i}(m \times n)$. Each line of such vector respectively collects the exergy equivalent to the amount of primary coal, oil and natural gas endogenously produced by each sector of the ith economy.

$$\mathbf{R_c} = [\mathbf{R_1} \quad \ldots \quad \mathbf{R_i} \quad \ldots \quad \mathbf{R_R}] \quad (4.21)$$

Specific and total exergy embodied in products of the total final demand, namely $\mathbf{e_c}(n_{tot} \times m)$ and $\mathbf{E_c}(n_{tot} \times m)$, can be finally derived by applying the Leontief model to the defined MRIO table as in relations (4.22) and (4.23), where $\mathbf{I_c}(n_{tot} \times n_{tot})$ is an *identity matrix* of appropriate dimensions.

$$\mathbf{B_c} = \mathbf{R_c \hat{x}_c^{-1}}; \quad \mathbf{A_c} = \mathbf{Z_c \hat{x}_c^{-1}} \quad (4.22)$$

$$\mathbf{L_c} = (\mathbf{I_W} - \mathbf{A_c})^{-1} \rightarrow \begin{cases} \mathbf{e_c} = (\mathbf{B_c L_c})^T \\ \mathbf{E_c} = diag(\mathbf{f_c i_f})\, \mathbf{e_c} \end{cases} \quad (4.23)$$

Notice that even if *model c* returns specific and total embodied exergy of national production for all the World's economies, only results of the jth country under investigation should be considered.

Model d—Multi-regional model with bi-directional trades

This model is introduced by the Author as a modified version of the uni-directional *model c*. According to this model, the amount and the direction of both the *imported* and *exported* products of the jth national economy under investigation are identified, while for the other $R - 1$ economies the remaining trades are endogenized in the domestic production in a similar way as for *model b*. The total production balance can be thus written as in relation (4.24).

$$\mathbf{x_d} = \mathbf{Z_d i_Z}(n_{tot} \times 1) + \mathbf{f_d i_f}(R \times 1)$$

$$
\begin{bmatrix} \mathbf{x_1} \\ \mathbf{x_j} \\ \mathbf{x_R} \end{bmatrix} = \begin{bmatrix} \mathbf{Z_{11}} & \mathbf{Z_{1j}} & - \\ \mathbf{Z_{j1}} & \mathbf{Z_{jj}} & \mathbf{Z_{jR}} \\ - & \mathbf{Z_{Rj}} & \mathbf{Z_{RR}} \end{bmatrix} \mathbf{i_Z} + \begin{bmatrix} \mathbf{f_{11}} & (\mathbf{ex_{1j}} - \mathbf{Z_{1j}i}) & - \\ (\mathbf{ex_{j1}} - \mathbf{Z_{j1}i}) & \mathbf{f_{jj}} & (\mathbf{ex_{jR}} - \mathbf{Z_{jR}i}) \\ - & (\mathbf{ex_{Rj}} - \mathbf{Z_{Rj}i}) & \mathbf{f_{RR}} \end{bmatrix} \mathbf{i_f}
$$

$$(4.24)$$

The World exogenous transaction matrix $\mathbf{R_d}(m \times n_{tot})$ is defined in the same way as for *model c*. Exergy embodied in the final demand is accounted for thanks to relations (4.25) and (4.26). As for the *model c*, only results of the *j*th country under investigation must be considered.

$$\mathbf{B_d} = \mathbf{R_d \hat{x}_d^{-1}}; \quad \mathbf{A_d} = \mathbf{Z_d \hat{x}_d^{-1}} \tag{4.25}$$

$$\mathbf{L_d} = (\mathbf{I_W} - \mathbf{A_d})^{-1} \rightarrow \begin{cases} \mathbf{e_d} = (\mathbf{B_d L_d})^{\mathrm{T}} \\ \mathbf{E_d} = \mathrm{diag}(\mathbf{f_d i_f}) \, \mathbf{e_d} \end{cases} \tag{4.26}$$

Model e—Multi-regional model with multi-directional trades
A greater detail is finally required for the application of *model e*: indeed, national production balance (4.27) must be written for all the R national economies as in relation (4.27), by defining destination and amount of international trades of products for each *i*th national economy as part of intermediate consumption or final demand of the other $R - 1$ economies. This method allows to remove all the hypotheses related to the treatment of international trades previously introduced: thus, it provides a formally correct evaluation of the exergy embodied in final demand produced by national economies. The World production balance (4.28) can be written by collecting all the detailed national production balances (4.27), and the structure of MRIO table for such model is represented in Fig. 4.6.

$$\mathbf{x_i} = \mathbf{Z_{ii}i}(n \times 1) + \mathbf{f_{ii}} + \sum_{j=1}^{R} \left[\mathbf{Z_{ij}i}(n \times 1) + \mathbf{f_{ij}}\right]_{j \neq i} \tag{4.27}$$

$$\mathbf{x_e} = \mathbf{Z_e i_Z}(n_{tot} \times 1) + \mathbf{f_e i_f}(R \times 1)$$

$$
\begin{bmatrix} \mathbf{x_1} \\ \mathbf{x_j} \\ \mathbf{x_R} \end{bmatrix} = \begin{bmatrix} \mathbf{Z_{11}} & \mathbf{Z_{1j}} & \mathbf{Z_{1R}} \\ \mathbf{Z_{j1}} & \mathbf{Z_{jj}} & \mathbf{Z_{jR}} \\ \mathbf{Z_{R1}} & \mathbf{Z_{Rj}} & \mathbf{Z_{RR}} \end{bmatrix} \mathbf{i_Z} + \begin{bmatrix} \mathbf{f_{11}} & \mathbf{f_{1j}} & \mathbf{f_{1R}} \\ \mathbf{f_{j1}} & \mathbf{f_{jj}} & \mathbf{f_{jR}} \\ \mathbf{f_{R1}} & \mathbf{f_{Rj}} & \mathbf{f_{RR}} \end{bmatrix} \mathbf{i_f} \tag{4.28}
$$

Again, the World exogenous transactions matrix $\mathbf{R_e}(m \times n_{tot})$ is defined in the same way as for *models c* and *d*, and the evaluation of embodied exergy in final

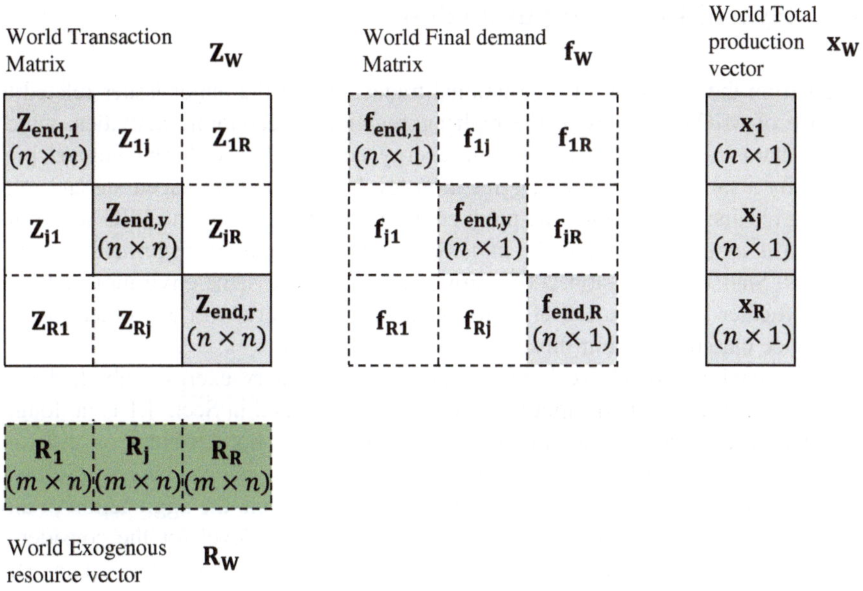

World Transaction Matrix $\mathbf{Z_W}$

$\mathbf{Z_{end,1}}$ $(n \times n)$	$\mathbf{Z_{1j}}$	$\mathbf{Z_{1R}}$
$\mathbf{Z_{j1}}$	$\mathbf{Z_{end,y}}$ $(n \times n)$	$\mathbf{Z_{jR}}$
$\mathbf{Z_{R1}}$	$\mathbf{Z_{Rj}}$	$\mathbf{Z_{end,r}}$ $(n \times n)$

World Final demand Matrix $\mathbf{f_W}$

$\mathbf{f_{end,1}}$ $(n \times 1)$	$\mathbf{f_{1j}}$	$\mathbf{f_{1R}}$
$\mathbf{f_{j1}}$	$\mathbf{f_{end,y}}$ $(n \times 1)$	$\mathbf{f_{jR}}$
$\mathbf{f_{R1}}$	$\mathbf{f_{Rj}}$	$\mathbf{f_{end,R}}$ $(n \times 1)$

World Total production vector $\mathbf{x_W}$

$\mathbf{x_1}$ $(n \times 1)$
$\mathbf{x_j}$ $(n \times 1)$
$\mathbf{x_R}$ $(n \times 1)$

$\mathbf{R_1}$ $(m \times n)$	$\mathbf{R_j}$ $(m \times n)$	$\mathbf{R_R}$ $(m \times n)$

World Exogenous resource vector $\mathbf{R_W}$

Fig. 4.6 Structure of the multi-regional input-output table according to model e

demand products of all the R national economies is performed through relations (4.29) and (4.30).

$$\mathbf{B_e} = \mathbf{R_e}\hat{\mathbf{x}}_e^{-1}; \quad \mathbf{A_e} = \mathbf{Z_e}\hat{\mathbf{x}}_e^{-1} \tag{4.29}$$

$$\mathbf{L_e} = (\mathbf{I_W} - \mathbf{A_e})^{-1} \rightarrow \begin{cases} \mathbf{e_e} = (\mathbf{B_e}\mathbf{L_e})^{\mathrm{T}} \\ \mathbf{E_e} = \mathrm{diag}(\mathbf{f_e}\mathbf{i_f})\,\mathbf{e_e} \end{cases} \tag{4.30}$$

If all the m kind of primary non-renewable resources collected by the exogenous transactions matrix $\mathbf{R_e}$ are measured in homogeneous units (e.g. *toe*, *J*, etc.), the total embodied exergy in the World final demand equals the total amount of primary exergy absorbed by the World economy, as in relation (4.31).

$$\left. \begin{array}{l} \mathbf{i}(1 \times m) \cdot \mathbf{R_e}(m \times n_{tot}) \cdot \mathbf{i}(n_{tot} \times 1) = R_{e,tot} \\ \mathbf{i}(1 \times m) \cdot [\mathbf{i}(1 \times n_{tot}) \cdot \mathbf{E_e}(n_{tot} \times m)]^{\mathrm{T}} = E_{e,tot} \end{array} \right\} \quad R_{W,tot} = E_{e,tot} \tag{4.31}$$

This equality is not respected for all the other previously introduced models, for which the difference between total consumption of resources and total embodied exergy of production increases as the share of international trades increases with respect to the total endogenous production. Because of its features and relevance, *model e* will be assumed as the reference in the following.

4.2   Hybrid Input-Output analysis

Apart from the treatment of international trades, one of the major issues related to the use of MIOTs resides in the high aggregation of economic activities, which makes hard to distinguish among the exergy embodied in different outputs produced by a same sector. Such aggregation level may produce inaccurate and misleading results: for example, the production of 1 € of values by the ISIC sector no. 3110 *Manufacture of electric motors, generators and transformers* (United Nations. Statistical Division 2008) turns out to have the same environmental burdens, whether it refers to 1 € of motor, generator, transformer or any other product or service classified within this sector of the economy.

If the analyst's objective is to account for the primary exergy embodied in *a specific product*, standard Input-Output analysis described in Sect. 4.1 is no longer adequate (Suh and Nakamura 2007). To overcome this problem, a different approach based on the so-called *Hybrid Input-Output analysis* can be used (Suh and Huppes 2005; Suh et al. 2004). According to this approach, the aggregation of any sector of the economy is selectively gained to adequate level for the considered product, using detailed process specific information, while the supply chains still covers the entire economy represented by the national MIOTs (Nakamura et al. 2007; Suh and Nakamura 2007). Notice that the term *hybrid* is usually adopted by the literature to indicate that such method relies on the joint application of both *Process Analysis* and *Input-Output analysis* (see Chap. 2) (Hendrickson et al. 2010; Suh and Huppes 2005). However, since mathematical equivalency between PA and IOA has been demonstrated in Sect. 2.4.2, the term *hybrid* denotes here the employed data sources: supply chains are modelled through the MIOTs, while specific survey are required to characterize the detailed system under investigation (Hendrickson et al. 2010; Suh and Huppes 2005).

The Hybrid Input-Output analysis is here introduced and formalized in a similar way as the Hybrid models suggested by (Suh and Huppes 2005; Hendrickson et al. 1998; Joshi 1999) and (Ferrão and Nhambiu 2009).

4.2.1   General formulation of the Hybrid Input-Output analysis

To account for primary exergy invoked for the production of detailed goods or services by a generic economic system, it is necessary to apply the Leontief model defined in Sect. 2.3.2 to the Hybrid Input-Output system represented in Fig. 4.7: the latter is composed by the supply chains, modelled through national or international MIOTs (*background system*, in grey), linked to the IOT of the analyzed system for which a detailed analysis is required (*foreground system*, in yellow).

In other words, any national economy, represented by a MIOT, can always be expressed as a hybrid productive system *H* composed by the same *National*

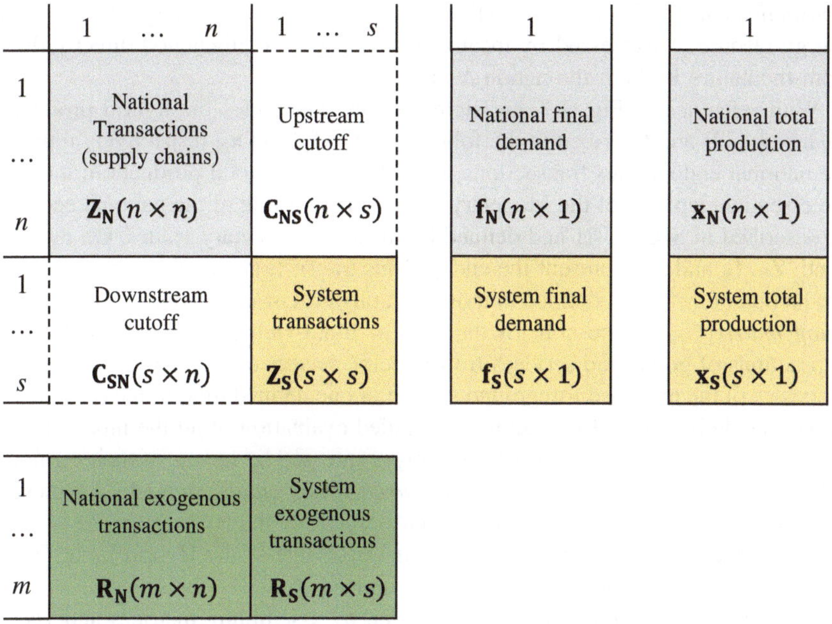

	1 ... n	1 ... s	1	1
1 ... n	National Transactions (supply chains) $\mathbf{Z_N}(n \times n)$	Upstream cutoff $\mathbf{C_{NS}}(n \times s)$	National final demand $\mathbf{f_N}(n \times 1)$	National total production $\mathbf{x_N}(n \times 1)$
1 ... s	Downstream cutoff $\mathbf{C_{SN}}(s \times n)$	System transactions $\mathbf{Z_S}(s \times s)$	System final demand $\mathbf{f_S}(s \times 1)$	System total production $\mathbf{x_S}(s \times 1)$
1 ... m	National exogenous transactions $\mathbf{R_N}(m \times n)$	System exogenous transactions $\mathbf{R_S}(m \times s)$		

Fig. 4.7 Structure of a generic hybrid input output table. The detailed system composed by s processes is operating in a national system of n economic activities. Both the system and the national supply chains may interact with the natural environment

economy N and by the *System S*, operating within one *j*th producing sector of the national economy *N*. Therefore, for a given time period, usually a year, the total production balance of the Hybrid system *H* can be expressed by relations (4.32) and (4.33).[1]

$$\mathbf{x_H} = \mathbf{Z_H i} + \mathbf{f_H}$$

$$\begin{bmatrix} \mathbf{x_N} \\ \mathbf{x_S} \end{bmatrix} = \begin{bmatrix} \mathbf{Z_N} & \mathbf{C_{NS}} \\ \mathbf{C_{SN}} & \mathbf{Z_S} \end{bmatrix} \cdot \mathbf{i}[(n+s) \times 1] + \begin{bmatrix} \mathbf{f_N} \\ \mathbf{f_S} \end{bmatrix} \tag{4.32}$$

$$\mathbf{R_H} = [\mathbf{R_N} | \mathbf{R_S}] \tag{4.33}$$

Matrix $\mathbf{Z_H}[(n+s) \times (n+s)]$ is the *hybrid transactions matrix*, while $\mathbf{f_H}[(n+s) \times 1]$ and $\mathbf{x_H}[(n+s) \times 1]$ are respectively the *hybrid final demand and total production vectors*. Notice that the above introduced hybrid matrix and vectors may be defined in mixed units: indeed, IOT of the foreground system *S* could be defined in physical, monetary or even mixed units. Finally, the *hybrid exogenous*

---

[1]Notice that the national final demand $\mathbf{f_N}$ is expressed here as one single column vector calculated through relation (4.15).

transactions matrix $\mathbf{R_H}[m \times (n+s)]$ collects the amount of primary non-renewable energy-resources (expressed by means of their exergy equivalents) directly drawn from the nature by both the nation N and the system S.

With reference to Fig. 4.7, the elements that constitutes the hybrid production balance (4.32) are described in the following. $\mathbf{Z_N}$, $\mathbf{f_N}$ and $\mathbf{x_N}$ respectively represent the national endogenous transactions, final demand and total production: these are the essential elements of the Monetary Input-Output Table of the national economy N, described in Sect. 4.1.1 and defined by means of monetary values. On the other hand, $\mathbf{Z_S}$, $\mathbf{f_S}$ and $\mathbf{x_S}$ represent the endogenous transactions, final demand and total production of system S, defined in monetary, physical or mixed units. The *upstream cutoff matrix* $\mathbf{C_{NS}}$ is the core of the Hybrid Input-Output model: it collects the transactions of goods and services from specific sectors of the economy to specific processes of the analyzed foreground system in one defined time frame (usually one year). The definition of $\mathbf{C_{NS}}$ requires a detailed evaluation of all the inputs that the System receives from the national economy, expressed by means of their economic values. Conversely, each element of the *downstream cutoff matrix* $\mathbf{C_{SN}}$ represents the amount of product flowing from each process of the System to one or more sectors of the National economy. Notice that each line of the Hybrid Input-Output system must be compiled in homogeneous units.

Once the Hybrid Input-Output table, given by assembling matrices and vectors $\mathbf{Z_H}$, $\mathbf{f_H}$, $\mathbf{x_H}$ and $\mathbf{R_H}$, has been completely characterized, embodied exergy in average products of both the nation N and detailed products of the considered system S can be calculated thanks to the application of Leontief model through relations (4.34) and (4.35).

$$\mathbf{A_H} = \mathbf{Z_H}\hat{\mathbf{x}}_H^{-1}; \quad \mathbf{B_H} = \mathbf{R_H}\hat{\mathbf{x}}_H^{-1}; \quad \mathbf{L_H} = (\mathbf{I_H} - \mathbf{A_H})^{-1} \qquad (4.34)$$

$$\mathbf{e_H} = \begin{bmatrix} \mathbf{e}_{H,N}(n \times m) \\ \hline \mathbf{e}_{H,S}(s \times m) \end{bmatrix} = (\mathbf{B_H}\mathbf{L_H})^T; \quad \mathbf{E_H} = \begin{bmatrix} \mathbf{E}_{H,N}(n \times m) \\ \hline \mathbf{E}_{H,S}(s \times m) \end{bmatrix} = \hat{\mathbf{f}}_H \mathbf{e_H} \qquad (4.35)$$

For the sake of simplicity, the foreground system here defined by the national economy N has been considered as closed towards international trades of goods and services. However, this hypothesis can be removed by applying one of the international trades models introduced in Sect. 4.1 to the hybrid system above defined.

The Hybrid system of Fig. 4.7 may be defined for two main purposes:

- *Increasing the accuracy of Input-Output analysis.* The Hybrid model can be used to analyze large productive systems, supply chains or even regions operating within a larger economic system. In this case, the Hybrid model is used to increase the accuracy of ExIO analysis by increasing the detail of relevant portions of the economy represented by the national MIOT;
- *Life Cycle Assessment of detailed systems.* In case of new or existing systems operating in a defined economic sector within a larger national economy, the Hybrid model should be properly defined in order to cover all the life cycle phases of the analyzed system.

4.2.2 Increasing the accuracy of Input-Output analysis

If the total production of system S is non-negligible compared to the total production of the national economic sector in which it operates, the application of the Leontief model to the hybrid system H provides two main benefits: (1) it allows to account for the primary exergy embodied in products of that system S; (2) it increases the quality and the accuracy of the results for all the other products of the national economy N. As an instance, operation of natural gas power plants could be decoupled from its broader ISIC sector D351 "*Electric power generation, transmission and distribution*": due to the increase in the detail of the defined hybrid system and the relevance of natural gas power generation within the national energy generation sector, application of Leontief model leads to compute the primary exergy requirements of natural gas power plants and also to increase the accuracy of results of all the other national products.

In this case, one fundamental issue in defining the Hybrid system is that all the transactions of the foreground system S must be collected according to the purpose that such products actually have in the national economy N. This means that products of system S delivered to the household will be collected in the final demand vector $\mathbf{f_S}$, while all the other endogenous transactions will be collected in the downstream cutoff matrix $\mathbf{C_{SN}}$.

Moreover, since the foreground system S operates *inside* the national economy N, relations (4.32) and (4.33) must represent the total production and the total exogenous transactions of the national economy N, defined by relation (4.14). Therefore, national and system transactions, as well as the exogenous transactions, must be *adjusted* in order to avoid double counting errors: in other words, the system S have to be numerically *extracted* from the national economy N. Once the jth sector of the national economy N in which the system S operates is known and the system IOT and the upstream cutoff matrix $\mathbf{C_{NS}}$ have been characterized, the following adjustments should be performed:

- The *adjusted National transaction matrix* $\overline{\mathbf{Z}}_N$ can be derived by subtracting the total products absorbed by S (column sum of matrix $\mathbf{C_{NS}}$) from the intermediate inputs of the jth sector of N ($\mathbf{Z}_{N,c:j}$ indicates the jth column of $\mathbf{Z_N}$), as expressed by relation (4.36).

$$\overline{\mathbf{Z}}_{N,c:j} = \mathbf{Z}_{N,c:j}(n \times 1) - [\mathbf{C_{NS}}(n \times s) \cdot \mathbf{i}(s \times 1)] \qquad (4.36)$$

- The *national adjusted final demand vector* $\overline{\mathbf{f}}_N$ results reducing the final demand of jth economic sector ($\mathbf{f}_{N,r:j}$ indicates the jth row of $\mathbf{f_N}$ matrix) by the total useful products of the analyzed system S, as in relation (4.37). Since the final demand vectors of N and S may be expressed in hybrid units, final demand of the system S has to be multiplied by the *average price vector* $\mathbf{p_S}(s \times 1)$ of the products of system S, converting the entire vector to monetary units;

$$\bar{\mathbf{f}}_{N,r:j} = \mathbf{f}_{N,r:j} - \{\mathbf{i}(1 \times s) \cdot [\hat{\mathbf{p}}_S(s \times s) \cdot \mathbf{f}_S(s \times 1)]\} \qquad (4.37)$$

- Evaluation of the *adjusted total production vector* is straightforward and is represented by relation (4.38). Since system S operates within the national economy N, the total intermediate production of the nation remains constant. However, total national production vector may change only if the system S delivers its products for final demand purposes;

$$\bar{\mathbf{x}}_N = \begin{bmatrix} \bar{\mathbf{Z}}_N & \mathbf{E}_{NS} \end{bmatrix} \cdot \mathbf{i}[(n+s) \times 1] + \bar{\mathbf{f}}_N \qquad (4.38)$$

- The *hybrid exogenous resources matrix* $\mathbf{R}_H[m \times (n+s)]$ is defined as in relation (4.39): if the system directly absorbs primary energy-resources, it is required to evaluate the adjusted exogenous resources matrix $\bar{\mathbf{R}}_H$ by subtracting the primary exergy resources absorbed by the system from the primary exergy resources absorbed by the jth economic sector the system operates within.

$$\mathbf{R}_H = [\bar{\mathbf{R}}_N | \mathbf{R}_S] \rightarrow \bar{\mathbf{R}}_{N,c:j} = \mathbf{R}_{N,c:j} - [\mathbf{R}_S(m \times s) \cdot \mathbf{i}(s \times 1)] \qquad (4.39)$$

Once the adjusted hybrid system has been defined, embodied exergy of average products of both the nation N and detailed products of the considered system S can be calculated thanks again to the application of Leontief model through relations (4.34) and (4.35).

4.2.3   Life Cycle Assessment of detailed systems

If the objective of the analyst consists in the evaluation of the primary exergy requirements invoked by the whole life cycle of a detailed system S, application of the Leontief model to the hybrid system differs with respect to the approach described in Sect. 4.2.2.

First of all, to perform a LCA of a generic system it is required to define both the object of the analysis (i.e. the *functional unit*) and the *time boundaries* of the considered system. To account for all the primary exergy requirements of the considered system, boundaries of the hybrid model should cover all the life cycle phases of the system (see Fig. 4.8). If a life cycle phase lasts more than one year, the analyst should carefully define the hybrid system using different MIOTs (if available) to characterize the background system (this usually happens with operation phase).

Another fundamental aspect concerns the definition of the foreground system. With respect to the approach described in Sect. 4.2.2, here the system final demand vector $\mathbf{f}_S$ only collects the useful products of the foreground system in each life cycle phase, while all the other by-products are collected in the upstream cutoff

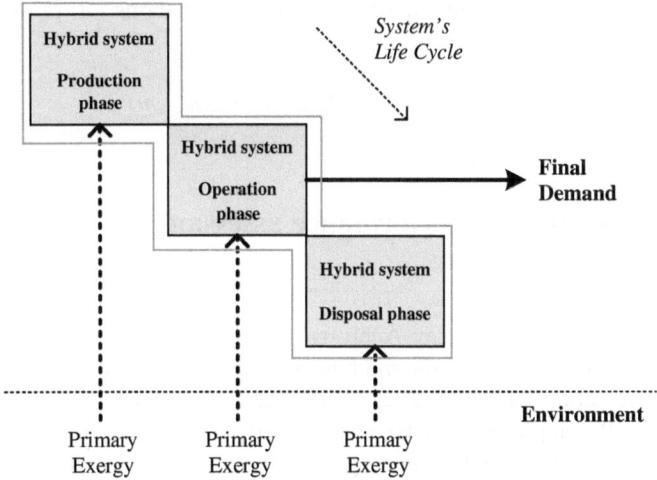

Fig. 4.8 Life cycle phases of a generic energy system. Notice that the system produces useful outputs only during its operation phase

matrix $\mathbf{C_{NS}}$, disregarding where such products are actually consumed. Indeed, based on the theoretical discussion of Sect. 2.3, Input-Output analysis allows only to account for the embodied exergy of final demand products. This approach was previously discussed in the literature with different names by many Authors: *Model II* (Joshi 1999), *Final Demand Approach* (Miller and Blair 2009b) and *Input—Output Hybrid method* (Heijungs and Suh 2002; Suh and Huppes 2005).

With respect to the approach described in Sect. 4.2.2, the definition of the hybrid system for the purpose of LCA of *new products* does not require the background system to be adjusted.

Once the hybrid system of each life cycle phase has been defined, exergy embodied in average products of both the nation N and detailed products of the considered system S can be calculated, thanks again to the application of Leontief model through relations (4.34) and (4.35). With reference to relation (4.40), the total primary exergy invoked by life cycle of the system is calculated as the sum of the primary exergy consumptions caused by each life cycle phase, while the specific exergy embodied in the final demand is obtained by a proper allocation of the total primary exergy on the final demand produced by the system during its operation phase (i.e. the functional unit).

$$Ex_H = Ex_{H,Construction} + yr \cdot Ex_{H,Operation} + Ex_{H,Disposal}; \quad ex_H = \frac{Ex_H}{\mathbf{i}(1 \times s)\mathbf{f}_{S,Operation}}$$

$$(4.40)$$

Notice that the total exergy embodied in the products of the system can be further decomposed according to relation (4.41), assessing the contribution of each single input delivered by the national economy in terms of primary exergy requirements.

$$\mathbf{E}_{NS}(n \times m) = diag[\mathbf{C}_{NS} \cdot \mathbf{i}] \cdot \mathbf{e}_N(n \times m) \tag{4.41}$$

4.3 Thermodynamic performance assessment of Energy Conversion Systems

As discussed in Chap. 3, Exergy Analysis (ExA) is widely adopted for the evaluation and optimization of the thermodynamic performances of energy conversion systems (Ayres et al. 1998; Kotas 2013; Moran and Sciubba 1994; Pavelka et al. 2015). Generally, application of ExA is focused on the quantification of thermodynamic irreversibilities within energy conversion systems, identifying possible margins for improvements and reduction of direct exergy requirements. However, boundaries of ExA are restricted to the physical layout of the considered system during its operative phase: primary exergy expenses indirectly caused by the production of non-energy-related goods and services required by the system during its whole life cycle are thus neglected (Kostowski et al. 2014; Lior and Zhang 2007).

Any proposed design improvement based on results of ExA should be verified in a broader perspective: *a reduction in the internal irreversibilities within a given system may not always be accompanied by a reduction in its primary resources requirement*. In this view, special attention should be devoted to renewable systems, the penetration of which in national production systems is continuously increasing (Colombo et al. 2014; Pina et al. 2013; Raugei et al. 2007, 2012). Starting from early 1970s, many different Thermodynamics-based methods have been proposed by the literature to evaluate the overall thermodynamic performances of energy conversion systems based on exergy (Ayres et al. 1996; Ayres et al. 1998; Cabeza et al. 2014; Cornelissen and Hirs 2002; Udo de Haes and Heijungs 2007). These methods can be generally collected in two main groups: *Exergy Cost Theory* (ECT), *Exergy Life Cycle Assessment* (ELCA).

4.3.1 Relation between Exergy Cost Theory and Input-Output analysis

The Exergy Cost Theory (ECT) was proposed by Valero as a set of rules and theoretical tools to account for the exergy cost of products of energy conversion systems, defined as the total amount of exergy required by the system to produce its products (Lozano and Valero 1993; Torres Cuadra et al. 2008; Valero et al. 1986a, b; Valero and Torres Cuadra 2006). Moreover, once all the interactions among components are characterized by means of exergy, the ECT analysis allows to understand the cost formation structure of the products, quantifying the relevance that internal irreversibilities have in increasing such costs, and defining criteria for optimization and diagnosis purposes (Lozano et al. 1994; Valero et al. 1994). Because of its features, the application of the ECT is usually limited to the boundaries of the considered energy system: this is due to the fact that the operation of supply chains involves interactions of goods and services that cannot always be characterized by means of exergy.

Given a generic energy conversion system S formed by s components connected to each other and with the environment by transactions of products quantified by means of exergy, ECT can be generally applied as follows. All the exergy flows are classified according to the Resource-Product-Loss (RPL) criterion, based on their purpose. Productive components generate useful products, whereas dissipative components are responsible for the disposal of any residues created during the process (condensers, filters, stacks, etc.) (Lazzaretto and Tsatsaronis 2006; Querol et al. 2012a). Given one defined time frame, e.g. one year of system operation, the exergy balance and exergy efficiency can be defined for each ith component of the system as in relation (4.42): the amount of exergy resources equals the sum of products, losses and exergy destructions.

$$Ex_{R,i} = Ex_{P,i} + Ex_{L,i} + Ex_{D,i} \rightarrow \eta_{ex,i} = Ex_{P,i}/Ex_{R,i} \qquad (4.42)$$

In line with traditional formulation of Input-Output analysis presented in Sect. 2. 3 (Suh 2005; Suh and Huppes 2005), exergy production of all the components can be collected in the matrix form, as in relation (4.43). The amount of exergy produced by each component is collected in the *total production vector* $\mathbf{x_S}(s \times 1)$. System *endogenous transactions matrix* $\mathbf{Z_S}(s \times s)$ represents the amount of products of the ith component fueled as a resource to all the other jth components (Ex_{ij}), while the *final demand vector* $\mathbf{f_S}(s \times 1)$ collects the amount of exergy delivered outside the physical boundaries of the system as the useful product. Finally, the system *exogenous transactions vector* $\mathbf{R_S}(m \times s)$ collects the amount of exergy directly invoked by each component from outside the system boundaries. The Resources-*Products table* is here defined as the assembly of matrices and vectors $\mathbf{Z_S}$, $\mathbf{f_S}$, $\mathbf{x_S}$ and $\mathbf{R_S}$.

$$\mathbf{x_S} = \mathbf{Z_S} \cdot \mathbf{i}(n \times 1) + \mathbf{f_S} \qquad (4.43)$$

The *technical coefficients matrix* $\mathbf{A_S}(s \times s)$ and the *input coefficients vector* $\mathbf{B_S}(m \times s)$ are defined through relation (4.44). Elements of $\mathbf{A_S}$ represent the amount of exergy that the jth component must receive from the ith component in order to produce one unit of its product. Similarly, the ith element of $\mathbf{B_S}$ represents the amount of exergy that must be provided to the ith component from outside the system boundary to produce one unit of its product.

$$\mathbf{A_S} = \mathbf{Z_S} \cdot \hat{\mathbf{x}}_\mathbf{S}^{-1}; \quad \mathbf{B_S} = \mathbf{R_S} \cdot \hat{\mathbf{x}}_\mathbf{S}^{-1} \tag{4.44}$$

Thanks to the above introduced definitions, it is possible to evaluate the specific and total exergy cost of the products of the system according to the Leontief model, where $\mathbf{e_S}(s \times m)$ is the specific exergy cost vector, $\mathbf{E_S}(s \times m)$ is the total exergy cost vector, and $\mathbf{L_S}(s \times s)$ is the *Leontief inverse coefficients matrix*.

$$\mathbf{L_S} = (\mathbf{I} - \mathbf{A_S})^{-1} \rightarrow \left\{ \begin{array}{c} \mathbf{e_S} = (\mathbf{B_S} \cdot \mathbf{L_S})^\mathrm{T} \\ \mathbf{E_S} = \hat{\mathbf{f}}_\mathbf{S} \cdot \mathbf{e_S} \end{array} \right. \tag{4.45}$$

Design evaluation and optimization of the considered system can be performed exploiting ECT routines in the light of the values that a specific set of parameters assume, with the aim to iteratively reduce the exergy cost of its products (Abusoglu and Kanoglu 2009). Beside the specific exergy cost of products (4.45) and the exergy destructions and losses $\mathbf{Ex_{S,(D+L)}}(s \times 1)$ (4.46), one other fundamental parameter is the *exergy cost of exergy destructions and losses* $\mathbf{C_{S,(D+L)}}(s \times 1)$ (4.47), that reveals the impact of thermodynamic irreversibilities in terms of additional exogenous exergy requirements of each component (Querol et al. 2012b). Optimization process is performed by iterative changes in the design configuration of the system, focusing on the *i*th component characterized by the highest values of the exergy cost of exergy destructions and losses.

$$\mathbf{Ex_{S,(D+L)}} = [\mathbf{i}(1 \times s) \cdot \mathbf{Z_S}]^\mathrm{T} - \mathbf{Z_S} \cdot \mathbf{i}(s \times 1) \tag{4.46}$$

$$\mathbf{C_{S,(D+L)}} = \hat{\mathbf{Ex}}_{\mathbf{S,(D+L)}} \cdot \mathbf{c_S} \tag{4.47}$$

Iterative optimization of a system performed according to the above introduced indicators allows to minimize the direct exergy requirements of the products of the system during its operative phase (i.e. minimize internal irreversibilities).

Further details about the RPL classification, reallocation of the exergy cost of residues and other issues related to the application of Exergy Cost Theory can be found in literature (Agudelo et al. 2012). Based on the present discussion, ECT can be alternatively defined according to the same mathematical structure of Input-Output analysis, which appears to be a common cost accounting framework in both Industrial Ecology and Thermodynamics disciplines. Therefore, exergy can be efficiently used within the Input-Output framework to study the process of exergy cost formation of products, and this provides a proper mathematical structure for the ELCA model presented below.

4.3.2 Design Evaluation of Energy Conversion Systems based on ELCA

Besides ECT, much effort has been devoted in the joint application of traditional Exergy Analysis and Life Cycle Assessment techniques, defining the so-called *Exergy Life Cycle Assessment* (ELCA) framework (see Sect. 3.1.3) (Ayres et al. 1998; Cornelissen 1997; Liao et al. 2012).

The practical application of ELCA methodologies usually relies on existing standardized LCA tools and databases (Dones et al. 2007; Frischknecht and Rebitzer 2005; Hedemann et al. 2007). According to literature, Environmentally extended Input-Output analysis is recognized as the computational structure of LCA models and environmental accountings in general (Heijungs and Suh 2002, 2006; Suh 2009): for this reason, both ECT and ELCA can be efficiently reformulated using the Input-Output mathematics (Keshavarzian et al. 2016; Valero et al. 2010). Apart from Valero, only isolated attempts have been made to incorporate Input-Output techniques within the exergy cost accounting (Hau and Bakshi 2004; Ukidwe et al. 2009), and a unique formalization of ELCA methods based on Input-Output analysis is still missing.

Design evaluation and optimization of energy conversion systems are usually performed according to Exergy Cost Theory: different design configurations of the system are proposed, with the aim to minimize the exergy cost of products of the system, that is, to minimize the direct consumption of exergy of the system. However, according to this approach, the additional exergy requirements due to the operation of supply chains and to the other life cycle phases of the system are not accounted. To overcome this issue, ELCA analysis can be used to support the optimization process: it can be applied by extending time and space boundaries of ECT to encompass the supply chains that feed the analyzed system during its complete life cycle. In this case, ELCA can be applied according to the same rules and mathematics of Hybrid Input-Output analysis described in Sect. 4.2. This allows to define the following indicators, useful to stress whether any design improvement suggested by Exergy Cost Theory provides overall benefits in terms of primary non-renewable resources displacement:

- *Primary exergy cost of system products* (e_H, E_H). It represents the primary exergy embodied in system products over the whole life cycle of the system. It can be expressed either in absolute units (E_H, in J) or with respect to the unit of final demand of the system Ex_P (e_H, in J/J).

$$E_H = E_{H,Construction} + yr \cdot E_{H,Operation} + E_{H,Disposal}; \quad e_H = \frac{E_H}{yr \cdot Ex_P} \quad (4.48)$$

Due to its general definition, values of specific primary exergy cost of system products can be efficiently adopted to perform comparison of the overall energy-conversion performance of different technologies (e.g. specific consumption of primary-resources of small photovoltaic plant and large combined cycle can be coherently compared). In relation (4.48), yr and Ex_P respectively represent the previsioned years of the plant operative lifetime and the total exergy of the products of the system in one average year of operation;

- *Net embodied exergy requirements* ($E_{H,Net}$) is defined as the net amount of primary exergy required by the system over its whole Life Cycle, as shown by relation (4.49).

$$E_{H,Net} = E_{H,Construction} + yr \cdot \left(E_{H,Operation} - Ex_P \right) + E_{H,Disposal} \qquad (4.49)$$

$E_{H,Net}$ can be interpreted as the total embodied exergy E_H reduced by the total amount of products of the plant (expressed by means of exergy) throughout its whole operative lifetime. Notice that positive values will always result from the analysis of systems based on non-renewable resources, while small and even negative values may result for renewable systems;

- *Exergy Return on Investment* (*ExROI*) is defined as the ratio of the absolute value of Net embodied exergy requirements $E_{H,Net}$ and the embodied exergy related to the construction phase only, as shown by relation (4.50).

$$ExROI = \frac{\left| E_{H,Net} \right|}{E_{H,Construction}} \qquad (4.50)$$

ExROI can be defined only in case of negative values of $E_{H,Net}$: it quantifies how many times the primary (non-renewable) exergy investment borne in the construction phase is paid back the net exergy associated to the final demand production. This indicator has already been defined by the literature to assess the overall thermodynamic performances of renewable energy systems (Cleveland and Costanza 2008; Hall et al. 2014; Heun and de Wit 2012; Weißbach et al. 2013).

References

Abusoglu, A., & Kanoglu, M. (2009). Exergoeconomic analysis and optimization of combined heat and power production: A review. *Renewable and Sustainable Energy Reviews, 13*, 2295–2308.

Agudelo, A., Valero, A., & Torres, C. (2012). Allocation of waste cost in thermoeconomic analysis. *Energy, 45*, 634–643.

Ahmad, N., & Wyckoff, A. (2003). Carbon dioxide emissions embodied in international trade of goods.

Ayres, R. U., Ayres, L. W., Martinas, K. (1996). Eco-thermodynamics: Exergy and life cycle analysis 49.

Ayres, R. U., Ayres, L. W., & Martinás, K. (1998). Exergy, waste accounting, and life-cycle analysis. *Energy, 23*, 355–363.

Battjes, J. J., Noorman, K. J., & Biesiot, W. (1998). Assessing the energy intensities of imports. *Energy Economics, 20*, 67–83.

Baumol, W. J. (2000). Leontief's great leap forward: Beyond Quesnay, Marx and von Bortkiewicz. *Economic Systems Research, 12*, 141–152.

Bullard, C. W., Penner, P. S., & Pilati, D. A. (1978). Net energy analysis: Handbook for combining process and input-output analysis. *Resources and Energy, 1*, 267–313.

Bullard, C. W, I. I. I., Herendeen, R. A., Bullard, C. W., & Herendeen, R. A. (1975). The energy cost of goods and services. *Energy policy, 3*, 268–278.

Cabeza, L. F., Rincón, L., Vilariño, V., Pérez, G., & Castell, A. (2014). Life cycle assessment (LCA) and life cycle energy analysis (LCEA) of buildings and the building sector: A review. *Renewable and Sustainable Energy Reviews, 29*, 394–416.

Chenery, H. B. (1953). The structure and growth of the Italian economy.

Cleveland, C. J., & Costanza, R. (2008). Energy return on investment (EROI). *Encyclopedia of Earth (online), April*.

Colombo, E., Rocco, M. V., Toro, C., & Sciubba, E. (2014). An exergy-based approach to the joint economic and environmental impact assessment of possible photovoltaic scenarios: A case study at a regional level in Italy. *Ecological Modelling*.

Cornelissen, R. L. (1997). Thermodynamics and sustainable development; The use of exergy analysis and the reduction of irreversibility.

Cornelissen, R. L., & Hirs, G. G. (2002). The value of the exergetic life cycle assessment besides the LCA. *Energy Conversion and Management, 43*, 1417–1424.

Dietzenbacher, E., & Lahr, M. L. (2004). *Wassily Leontief and input-output economics*. Cambridge University Press Cambridge.

Dones, R., Bauer, C., Bolliger, R., Burger, B., Faist Emmenegger, M., Frischknecht, R., et al. (2007). Life cycle inventories of energy systems: results for current systems in Switzerland and other UCTE countries. *Ecoinvent report 5*.

Duchin, F., Lange, G.-M., Thonstad, K., Idenburg, A., & Cropper, M. L. (1996). The future of the environment: Ecological economics and technological change. *Journal of Economic Literature, 34*, 818–819.

Duchin, F., & Levine, S. H. (2013). Embodied resource flows in a global economy. *Journal of Industrial Ecology, 17*, 65–78.

Eurostat. (2008). *NACE rev.2. Statistical classification of economic activities in the European Community*. Luxembourg: Office for Official Publications of the European Communities.

Ferrão, P., & Nhambiu, J. (2009). A comparison between conventional LCA and hybrid EIO-LCA: Analyzing crystal giftware contribution to global warming potential. In S. Suh (Ed.), *Handbook of input-output economics in industrial ecology* (Vol. 23, pp. 219–230). Netherlands: Springer.

Frischknecht, R., & Rebitzer, G. (2005). The ecoinvent database system: A comprehensive web-based LCA database. *Journal of Cleaner Production, 13*, 1337–1343.

Gyftopoulos, E. P., & Beretta, G. P. (1991). *Thermodynamics: Foundations and applications*. Macmillian.

Hall, C. A. S., Lambert, J. G., & Balogh, S. B. (2014). EROI of different fuels and the implications for society. *Energy policy, 64*, 141–152.

Hau, J. L., & Bakshi, B. R. (2004). Expanding exergy analysis to account for ecosystem products and services. *Environmental Science and Technology, 38*, 3768–3777.

Hedemann, J., König, U., Cuche, A., Egli, N. (2007). Technical documentation of the ecoinvent database. *Final report ecoinvent data v2.2*.

Heijungs, R., & Suh, S. (2002). *The computational structure of life cycle assessment*. Springer.

Heijungs, R., & Suh, S. (2006). Reformulation of matrix-based LCI: From product balance to process balance. *Journal of Cleaner Production, 14*, 47–51.

Hendrickson, C., Horvath, A., Joshi, S., & Lave, L. (1998). Peer reviewed: Economic input-output models for environmental life-cycle assessment. *Environmental Science and Technology, 32*, 184A–191A.

Hendrickson, C. T., Horvath, A., Joshi, S., Klausner, M., Lave, L. B., & McMichael, F. C. (1997). Comparing two life cycle assessment approaches: a process model vs. economic input-output-based assessment, electronics and the environment, 1997. In *Proceedings of the 1997 IEEE international symposium on ISEE-1997* (pp. 176–181). IEEE.

Hendrickson, C. T., Lave, L. B., & Matthews, H. S. (2010). *Environmental life cycle assessment of goods and services: An input-output approach.* Routledge.

Heun, M. K., & de Wit, M. (2012). Energy return on (energy) invested (EROI), oil prices, and energy transitions. *Energy Policy, 40,* 147–158.

Isard, W. (1951). Interregional and regional input-output analysis: A model of a space-economy. *The Review of Economics and Statistics,* 318–328.

Isard, W. (1960). Methods of regional analysis.

Isard, W., & Langford, T. W. (1971). Regional input-output study: Recollections, reflections, and diverse notes on the Philadelphia experience.

Joshi, S. (1999). Product environmental life-cycle assessment using input-output techniques. *Journal of Industrial Ecology, 3,* 95–120.

Joshi, S. (2000). Life-cycle assessment using input-output. *Techniques, 3,* 95–120.

Kendrick, J. W. (1996). *The new system of national accounts.* Berlin: Springer.

Keshavarzian, S., Gardumi, F., Rocco, M. V., & Colombo, E. (2016). Off-design modeling of natural gas combined cycle power plants: An order reduction by means of thermoeconomic input-output analysis. *Entropy 18.*

Kostowski, W. J., Usón, S., Stanek, W., & Bargiel, P. (2014). Thermoecological cost of electricity production in the natural gas pressure reduction process. *Energy,* 1–9.

Kotas, T. J. (2012). *The exergy method of thermal plant analysis.* Paragon Publishing.

Kotas, T. J. (2013). *The exergy method of thermal plant analysis.* Elsevier.

Lazzaretto, A., & Tsatsaronis, G. (2006). SPECO: A systematic and general methodology for calculating efficiencies and costs in thermal systems. *Energy, 31,* 1257–1289.

Lenzen, M., Pade, L.-L., & Munksgaard, J. (2004). Multipliers in multi-region input-output models. *Economic Systems Research, 16,* 391–412.

Leontief, W., (1953). Studies in the structure of the American economy.

Leontief, W. (1974). Structure of the world economy: Outline of a simple input-output formulation. *The American Economic Review,* 823–834.

Leontief, W. (1986). *Input-output economics.* Oxford University Press.

Liao, W., Heijungs, R., & Huppes, G. (2012). Thermodynamic analysis of human–environment systems: A review focused on industrial ecology. *Ecological Modelling, 228,* 76–88.

Lior, N., & Zhang, N. (2007). Energy, exergy, and second law performance criteria. *Energy, 32,* 281–296.

Lozano, M. A., Bartolomé, J. L., Valero, A., & Reini, M. (1994). Thermoeconomic diagnosis of energy systems, *Flowers,* 149–156.

Lozano, M. A., & Valero, A. (1993). Theory of the exergetic cost. *Energy, 18,* 939–960.

Matthews, H., & Small, M. (2000). Extending the boundaries of life-cycle assessment through environmental economic input-output models. *Journal of Industrial Ecology, 4,* 7–10.

Miller, R. E., & Blair, P. D. (2009a). Input-output analysis: Foundations and extensions.

Miller, R. E., & Blair, P. D. (2009b). *Input-output analysis: Foundations and extensions.* Cambridge University Press: Cambridge.

Moran, M., & Sciubba, E. (1994). Exergy analysis: Principles and practice. *ASME Transactions Journal of Engineering Gas Turbines and Power, 116,* 285–290.

Moses, L. N. (1955). The stability of interregional trading patterns and input-output analysis. *The American Economic Review,* 803–826.

Nakamura, S., Nakajima, K., Kondo, Y., & Nagasaka, T. (2007). The waste input-output approach to materials flow analysis. *Journal of Industrial Ecology, 11,* 50–63.

Nations, U., Commission, E., Fund, I. M., Co-operation, O. F. E., & Development, Bank, W. (2009). System of National Accounts 2008. UN.

Nijdam, D. S., Wilting, H. C., Goedkoop, M. J., & Madsen, J. (2005). Environmental load from Dutch private consumption: How much damage takes place abroad? *Journal of Industrial Ecology, 9,* 147–168.

Pavelka, M., Klika, V., Vágner, P., & Maršík, F. (2015). Generalization of exergy analysis. *Applied Energy, 137*, 158–172.

Peters, G., & Hertwich, E. (2005). The global dimensions of Norwegian household consumption. *Journal of Industrial Ecology Submitted August* 31, 2005.

Peters, G. P., & Hertwich, E. G. (2006). The importance of imports for household environmental impacts. *Journal of Industrial Ecology, 10*, 89–109.

Pina, A., Silva, C. A., & Ferrão, P. (2013). High-resolution modeling framework for planning electricity systems with high penetration of renewables. *Applied Energy, 112*, 215–223.

Querol, E., González-Regueral, B., & Benedito, J. L. P. (2012a). Practical approach to exergy and thermoeconomic analyses of industrial processes.

Querol, E., Gonzalez-Regueral, B., & Perez-Benedito, J. L. (2012b). Practical approach to exergy and thermoeconomic analyses of industrial processes.

Raugei, M., Bargigli, S., & Ulgiati, S. (2007). Life cycle assessment and energy pay-back time of advanced photovoltaic modules: CdTe and CIS compared to poly-Si. *Energy, 32*, 1310–1318.

Raugei, M., Fullana-i-Palmer, P., & Fthenakis, V. (2012). The energy return on energy investment (EROI) of photovoltaics: Methodology and comparisons with fossil fuel life cycles. *Energy Policy, 45*, 576–582.

Round, J. I. (2001). Feedback effects in interregional input-output models: What have we learned?

Strømman, A. H., & Gauteplass, A., (2004). Domestic fractions of emissions in linked economies.

Suh, S. (2005). Theory of materials and energy flow analysis in ecology and economics. *Ecological Modelling, 189*, 251–269.

Suh, S. (2009). Handbook of input-output analysis economics in industrial ecology.

Suh, S., & Huppes, G. (2005). Methods for life cycle inventory of a product. *Journal of Cleaner Production, 13*, 687–697.

Suh, S., Lenzen, M., Treloar, G. J., Hondo, H., Horvath, A., Huppes, G., et al. (2004). System boundary selection in life-cycle inventories using hybrid approaches. *Environmental Science and Technology, 38*, 657–664.

Suh, S., & Nakamura, S. (2007). Five years in the area of input-output and hybrid LCA. *International Journal of Life Cycle Assessment, 12*, 351–352.

Torres Cuadra, C., Valero, A., Rangel, V., & Zaleta, A. (2008). On the cost formation process of the residues. *Energy, 33*, 144–152.

Treloar, G. J. (1997). Extracting embodied energy paths from input–output tables: Towards an input–output-based hybrid energy analysis method. *Economic Systems Research, 9*, 375–391.

Treloar, G. J. (1998). *Comprehensive embodied energy analysis framework*. Deakin University.

Tukker, A., Huppes, G., Oers, L.V., & Heijungs, R. (2006). Environmentally extended input-output tables and models for Europe.

Turner, K., Lenzen, M., Wiedmann, T., & Barrett, J. (2007). Examining the global environmental impact of regional consumption activities—part 1: A technical note on combining input–output and ecological footprint analysis. *Ecological Economics, 62*, 37–44.

Udo de Haes, H. A., & Heijungs, R. (2007). Life-cycle assessment for energy analysis and management. *Applied Energy, 84*, 817–827.

Ukidwe, N. U., Hau, J. L., & Bakshi, B. R., (2009). Thermodynamic input-output analysis of economic and ecological systems. In *Handbook of input-output economics in industrial ecology* (pp. 459–490). Springer.

United Nations. Statistical Division. (2008). *International Standard industrial classification of all economic activities (ISIC), Rev* (4th ed.). New York: United Nations.

Valero, A., Lozano, M. A., & Muñoz, M. (1986a). A general theory of exergy saving. Part 1: on the exergetic cost. In R. Gaggioli (Ed.), *Computer-aided engineering and energy systems. Second law analysis and modelling*. Zaragoza: The American Society of Mechanical Engineers.

Valero, A., Lozano, M. A., & Muñoz, M. (1986b). A general theory of exergy saving. Part 2: On the thermoeconomic cost. In R. Gaggioli (Ed.), *Computer-aided engineering and energy systems. Second law analysis and modelling*. Zaragoza: The American Society of Mechanical Engineers.

Valero, A., Lozano, M. A., Serra, L., & Torres, C. (1994). Application of the exergetic cost theory to the CGAM problem. *Energy, 19*, 365–381.

Valero, A., & Torres Cuadra, C. (2006). Thermoeconomic analysis. In *Exergy, energy system analysis and optimization*.

Valero, A., Usón, S., Torres, C., & Valero, A. (2010). Application of thermoeconomics to industrial ecology. *Entropy, 12*, 591–612.

Weber, C. L., & Matthews, H. S. (2008). Quantifying the global and distributional aspects of American household carbon footprint. *Ecological Economics, 66*, 379–391.

Weißbach, D., Ruprecht, G., Huke, A., Czerski, K., Gottlieb, S., & Hussein, A. (2013). Energy intensities, EROIs (energy returned on invested), and energy payback times of electricity generating power plants. *Energy, 52*, 210–221.

Wiedmann, T. (2009). A review of recent multi-region input-output models used for consumption-based emission and resource accounting. *Ecological Economics, 69*, 211–222.

Wiedmann, T., Lenzen, M., Turner, K., & Barrett, J. (2007). Examining the global environmental impact of regional consumption activities—Part 2: Review of input–output models for the assessment of environmental impacts embodied in trade. *Ecological Economics, 61*, 15–26.

Wiedmann, T., Wilting, H. C., Lenzen, M., Lutter, S., & Palm, V. (2011). Quo Vadis MRIO? Methodological, data and institutional requirements for multi-region input-output analysis. *Ecological Economics, 70*, 1937–1945.

Chapter 5
Internalization of human labour in Input-Output analysis

This chapter focuses on the internalization of *human labour* in the calculation of primary exergy embodied in goods and services. After a brief review about the main approaches proposed by literature, the *Bioeconomic Exergy based Input-Output analysis* (B-ExIO) is here introduced and discussed.

5.1 Internalization of human labour in LCA: literature review

According to the literature, quantitative evaluation of the environmental impact caused by human labour is one of the most controversial topics in Life Cycle Assessment (Sciubba 2004; Xu et al. 2009). Many disciplines face the study of human labour, ranging from economics to social sciences. Accounting for labour embodied in goods and services is a relevant practice in classical economics and social accountings: based on the *labor theory of value*, one of the main driver for economic value of goods and services consists in the amount of human labour (in the form of working hours) embodied in such products, as introduced by *Petty* (Dooley 2005). Both environmental and economic sciences are therefore involved in the debate concerning the treatment of human labour as a factor of production (Ayres 2004).

Human labour is defined by *economists* as an independent primary input with respect to the economic system: it is essential for the production of all the goods and services, but its creation lies outside the domain of the analysis (Faber and Manstetten 2009). Therefore, if the cost of labour is accounted as part of the production cost of the economy, this would result in a double accounting error. On the other hand, *Ecologists* recognize that workers are able to produce working hours only thanks to the depletion of natural resources. As a consequence, they claim to include resources and wastes embodied in working hours as additional

© The Author(s) 2016
M.V. Rocco, *Primary Exergy Cost of Goods and Services*,
PoliMI SpringerBriefs, DOI 10.1007/978-3-319-43656-2_5

environmental burdens of goods and services production. In support of this, *Ayres* states that working hours and capitals should be considered as intermediate rather than secondary inputs, because primary resources and waste emissions are required to produce both of them (Ayres 2001). Classic economic paradigms do not take into account for such contributions as part of the cost of production (i.e. clothing, housing, food, education and training and other workers' consumptions). For such reason, ecologists claim for inconsistency in the classic economic paradigm (Ayres 2004).

Activities devoted to the production of goods and services of *all* the economic systems actually require *direct* and *indirect* amounts of working hours in order to be productive. As a consequence, a certain amount of embodied working hours needs to be produced by workers that, in turn, spend their salaries buying and consuming products required to support their life (e.g. food, clothing, housing, etc.), and causing additional embodied environmental burdens that should be properly taken into account. The so-called *human labour internalization process* refers here to the inclusion of such effects within the environmental impacts of goods and services production (Ayres 2004; Pokrovskii 2011).

Practical attempts to internalize the effects of human labour in environmental and thermodynamic accountings can be found in literature, in the fields of Exergy and Emergy analyses. *Szargut* and colleagues have addressed the issue of human labour internalization in the definition of *Cumulative Exergy Consumption* (CExC) method (Szargut and Morris 1987; Szargut et al. 1987): according to a set of defined accounting rules, Szargut stated that including human labour in cumulative exergy accounting would result in a double accounting error (Szargut et al. 2002). Contrarily, the *Extended Exergy Accounting* (EEA), conceived by *Sciubba* (Sciubba 2001; Sciubba and Ulgiati 2005), defines a method to account for the exergy equivalent of human labour, and to include such and other contributions due to externalities into the overall primary exergy requirements of products. In brief, EEA accounts for primary exergy requirements of human labour postulating that the whole resources absorbed by any society is ultimately devoted to sustain workers who generate labour (Sciubba 2011). Based on this assumption, the embodied exergy of one working hour of the average worker can be evaluated as the ratio between the exergy absorbed by the society and the total number of working hours produced by the same society in the same period. A detailed review of EEA can be found in literature (Liao et al. 2012; Rocco et al. 2014).

Beside Extended Exergy Accounting, a comprehensive discussion related to the internalization of human labour can be found in the Emergy analysis literature (Brown and Herendeen 1996; Hau and Bakshi 2004): in particular, *Kamp* et al. reviewed, formalized and compared different possible approaches to internalize human labour within Emergy analysis (Kamp et al. 2016).

Other Authors tried to analyze the environmental effects of human labour production: *Costanza* proposes the use of a *Closed Input-Output model* to account for the embodied energy of products of the US economy, and to investigate the *energy theory of value* by comparing these results with the economic value of the same products (Costanza 1980; Costanza and Herendeen 1984). Some efforts along this line can be found also in *Duchin*'s publications (Duchin and Steenge 2007). In a different perspective, *Fukuda* have investigated the relation among labour production and primary exergy requirements at global scale (Fukuda 2003).

Many theoretical debates can be found in literature concerning the role of human labour in economy and ecology. However, only few Authors have focused on the issue of labour internalization from a methodological viewpoint, mainly due to the following reasons (Ayres 2004):

- Univocal and unambiguous definition of one *labor production process* is difficult and the unit process defined will be affected by great arbitrariness (Treloar 1998);
- Based on the literature, the contribution of human labour to the environmental impact of products of modern economic systems is expected to be negligible (Boustead and Hancock 1979; Consoli et al. 1993). However, numerical proof of this statement has not been provided yet.

In the following, the effects of human labour are internalized in primary energy-resources accountings through the Exergy-based Input-Output framework.

5.2   Bioeconomic ExIO analysis: theoretical definition

According to the Exergy-based Input-Output analysis, defined in Chap. 4, every national economy produces goods and services to fulfill the households' final demand. All the national economic activities are directly and indirectly involved in the production of such final demand, causing consumption of primary resources and emissions of wastes. With reference to Fig. 5.1, the production activities within the national economy define the so-called *circular economy* (Miller and Blair 2009): the final demand that sustains all the household activities is compensated by an opposite flow of human labour (i.e. working hours) that is invoked by the economic production. According to this model, human labour production lies outside the boundaries of the national production system: working hours thus result as like as any other exogenous transaction.

According to the *Neoclassical theory of Labor Supply* (King et al. 1988), the total final demand of goods and services is required by the households for two main purposes: (1) to sustain *labour production* and (2) to generate *leisure time production* (i.e. all other activities except for working hours production). In this perspective, and with reference to Fig. 5.2 (left side), households sector can be ideally divided into two distinct productive processes: the *human labour production process* requires a portion of the total final demand to sustain workers who generate labour; on the other hand, the *Leisure time production process* consists in a perfect

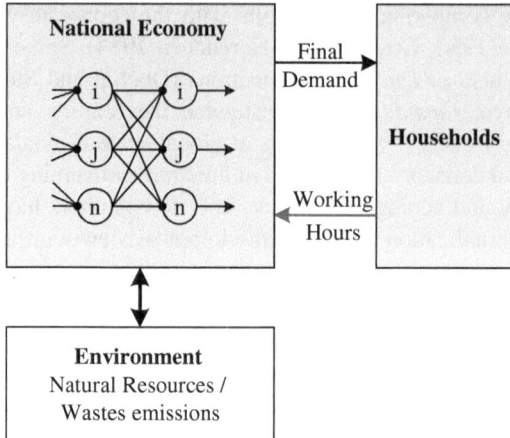

Fig. 5.1 Schematic outline of the standard input-output model that describes the circular nature of national economies

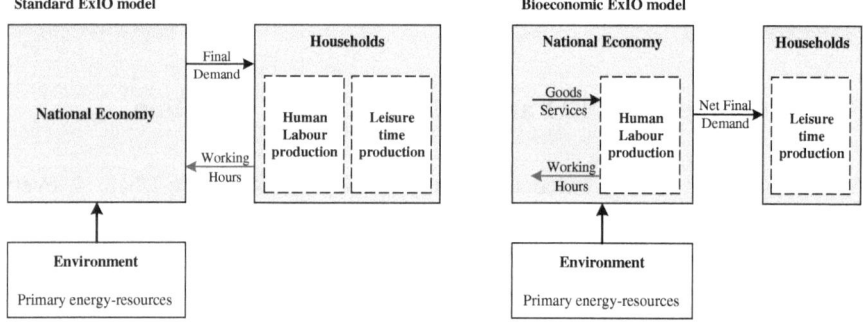

Fig. 5.2 Definition of the bioeconomic ExIO model (*right side*) starting from the standard IO model (*left side*)

sink, absorbing the remaining portion of the final demand to sustain all the leisure activities of the households.

The *Bioeconomic ExIO model* consists in a partially closed Input-Output model: it defines a quantitative criterion to share the final demand among labour and leisure activities, extending boundaries of conventional Input-Output model to internalize the *human labour production process* as a new production activity within the national economy. Such new sector of the economy receives a portion of goods and services invoked by the final demand and returns working hours as its unique product.

The term *Bioeconomic* explicitly refers to the *Bioeconomic paradigm*, proposed by *Nicholas Georgescu-Roegen* as an alternative to the *Neoclassical economic paradigm* (Georgescu-Roegen 1979; Mayumi 2009). The purpose of the economic production and the role of human labour were discussed by Georgescu-Roegen through the following example: *«[...] we should cure ourselves of what I have been calling "the circumdrome of the shaving machine", which is to shave oneself faster so as to have more time to work on a machine that shaves faster so as to have more time to work on a machine that shaves still faster, and so on ad infinitum. This change will call for a great deal of recanting on the part of all those professions which have lured man into this empty infinite regress. We must come to realize that an important prerequisite for a good life is a substantial amount of leisure spent in an intelligent manner»* (Georgescu-Roegen 1975). According to this example, Georgescu-Roegen claims that the role of human labour should be properly taken into account in the definition of the objective of the economic production. The Bioeconomic Input-Output model here proposed ought to be a small step forward in a proper internalization of human labour as an essential factor of the energy-resources metabolism of national economies. In the following, two fundamental elements for the definition of the Bioeconomic model are introduced: (1) the quantitative criterion required to share the total final demand among human labour production and leisure time production, and (2) the theoretical definition of the model as a *partially closed hybrid Input-Output model*.

5.2.1 Accounting for Products Required for Human Labour Production

With reference to relation (5.1), national final demand can be defined as the sum of four main contributions: *household purchases* (f_H), *purchases for (private) investment purposes* (f_I), *government purchases* (f_G) (federal, public, and local), and *exports* (f_E), each of one related to the ith productive sector of the economy (see Sect. 4.1.1).

$$\mathbf{f}(n \times 4) = \begin{bmatrix} \mathbf{f_H} & \mathbf{f_I} & \mathbf{f_G} & \mathbf{f_E} \end{bmatrix} \tag{5.1}$$

Distinguishing among the national final demand devoted to sustain production of human labour and leisure time is a complex task and it is affected by great uncertainty and arbitrariness due to the following reasons. First of all, products of the economy may be used by the households during both work and leisure activities (e.g. clothes, food, cars, etc.); secondly, many products are actually invoked in a time window greater than one year for investment purposes. Finally, there are different kind of workers that require different kind and amount of products (for instance, workers with higher income levels require larger amounts of goods and services compared to other workers with lower income levels.

To account for the national final demand related to human labour production, the Bioeconomic Input-Output model assumes the following fundamental hypotheses:

- The total final demand devoted to sustain households consumption $\mathbf{f_H}$ can be expressed as the sum of products devoted to sustain labour activities $\mathbf{f_{H,W}}$ and leisure activities $\mathbf{f_{H,L}}$, as represented by relation (5.2);

$$\mathbf{f_H}(n \times 1) = \mathbf{f_{H,W}} + \mathbf{f_{H,L}} \tag{5.2}$$

- With reference to the ith economic activity, the amount of products invoked by workers of one generic ith sector of the economy to generate labour is *proportional* to the amount of working hours absorbed by the same sector. In other words, the ratio between the final demand required from the ith sector $f_{H,W,i}$ and the total households final demand $f_{H,tot}$ equals the ratio between the amount of hours devoted to working activities $h_{W,i}$ and total hours lived by the entire population h_{tot}, as showed by relation (5.3).

$$\frac{f_{H,W,i}}{f_{H,tot}} = \frac{h_{W,i}}{h_{tot}} \quad \rightarrow \quad f_{W,i} = f_{H,tot} \cdot \frac{h_{W,i}}{h_{tot}} \tag{5.3}$$

The total final demand of households $f_{H,tot}$ and the total hours lived by the entire population h_{tot} can be respectively determined from relations (5.4) and (5.5), in which N_{pop} is the population of the considered country. Data related to the amount of working hours employed in each productive sector are collected by national bureaus of statistics or may be estimated according to the literature (Pyatt and Round 1985; Wales and Woodland 1977).

$$f_{H,tot} = \mathbf{i}(1 \times n) \cdot \mathbf{f_H}(n \times 1) \tag{5.4}$$

$$h_{tot} = N_{pop} \cdot 8760 \, h/y \tag{5.5}$$

Relation (5.3) can be written in the matrix form (5.6), by introducing the *working hours requirements vector* $\mathbf{h_W}(n \times 1)$, which collects the working hours required by each sector of the economy in one year.

$$\mathbf{f_{H,W}} = \mathbf{i}(1 \times n) \cdot \mathbf{f_H} \cdot \left(\frac{\mathbf{h_W}}{h_{tot}}\right) \tag{5.6}$$

With reference to Fig. 5.2, once the total final demand has been divided into the two aforementioned components according to relation (5.6), the human labour production process can be finally internalized within the economy.

5.2.2   Definition of the Bioeconomic ExIO model

The Bioeconomic ExIO model can be defined as a *Hybrid Input-Output model partially closed with respect to the households' final demand*. Let us consider the

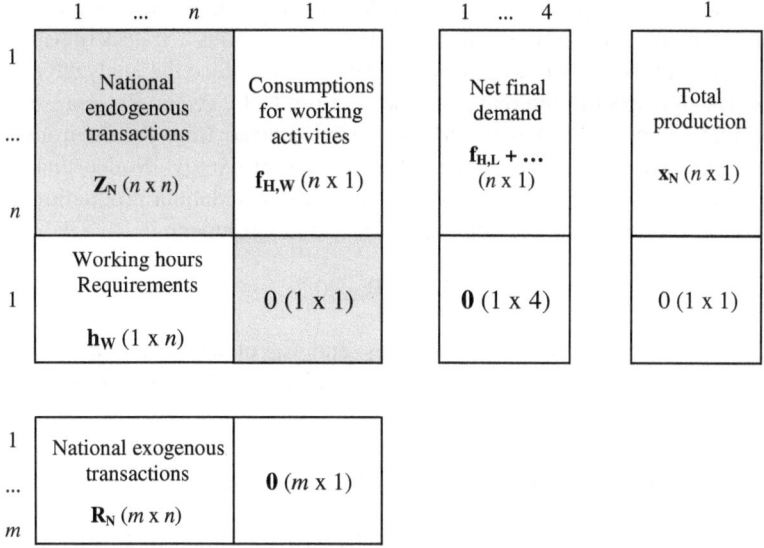

Fig. 5.3 Matrix definition of the bioeconomic input-output model

generic national economy N characterized by the national production balance (4.14) defined in Sect. 4.1.1. For the sake of simplicity, such economy has been considered here as closed towards international trades of goods and services; however, this hypothesis can be removed by applying one of the international trades models introduced in Sect. 4.1 to the hybrid system here defined.

With reference to Fig. 5.3, the national production balance (4.14) can be rewritten as in relation (5.7): vector $\mathbf{x_B}[(n+1) \times 1]$ is the total production vector, $\mathbf{Z_B}[(n+1) \times (n+1)]$ and $\mathbf{f_B}[(n+1) \times 4]$ are respectively the endogenous transactions and the final demand matrices. Notice that the above introduced vector and matrices are defined in mixed units: economic transactions are defined in monetary values for $\mathbf{Z_N}, \mathbf{x_N}, \mathbf{f_{H,I,G,E}}$, while products of the human labour production process are accounted in terms of working hours.

$$\mathbf{x_B} = \mathbf{Z_B}\mathbf{i} + \mathbf{f_B}$$

$$\begin{bmatrix} \mathbf{x_N} \\ h_{W,tot} \end{bmatrix} = \begin{bmatrix} \mathbf{Z_N} & \mathbf{f_{H,W}} \\ \mathbf{h_W^T} & - \end{bmatrix} \cdot \mathbf{i}[(n+1) \times 1] + \begin{bmatrix} \mathbf{f_{H,L}} + \mathbf{f_I} + \mathbf{f_G} + \mathbf{f_E} \\ - \end{bmatrix} \cdot \mathbf{i}(4 \times 1)$$

$$(5.7)$$

According to the above introduced national production balance (5.7), the human labour production sector produces working hours $\mathbf{h_W}(n \times 1)$ by consuming a portion of the final demand $\mathbf{f_{H,W}}(n \times 1)$ from the national economy. Notice that working hours are only produced for intermediate consumption of the economy, that is, they are not produced for final demand uses, nor self-consumed by the human labour production sector. The exogenous transactions matrix $\mathbf{R_B}[m \times (n+1)]$ is defined according to relation (5.8): since the analyzed economy is closed with respect to international trades of products, vector $\mathbf{R_B}$ represents only the amount of energy-resources *endogenously* extracted from the environment by each sector of the economy, expressed by means of exergy. Notice that the last column of $\mathbf{R_B}$ is zero, since it is assumed that the human labour production process does not absorb any resources directly from the environment.

$$\mathbf{R_B} = [\mathbf{R_{end}}|0] \tag{5.8}$$

Total and specific embodied exergy $\mathbf{e_B}$ and $\mathbf{E_B}$ of national final demand are finally derived through the application of Leontief model as in relations (5.9) and (5.10).

$$\mathbf{A_B} = \mathbf{Z_B} \cdot \hat{\mathbf{x}}_\mathbf{B}^{-1}; \quad \mathbf{B_B} = \mathbf{R_B} \cdot \hat{\mathbf{x}}_\mathbf{B}^{-1} \tag{5.9}$$

$$\mathbf{L_B}[(n+1) \times (n+1)] = \left(\mathbf{I} - \mathbf{A_B^T}\right)^{-1} \rightarrow \begin{cases} \mathbf{e_B}[(n+1) \times m] = (\mathbf{B_B L_B})^\mathbf{T} \\ \mathbf{E_B}[(n+1) \times m] = \hat{\mathbf{f}}_\mathbf{B} \mathbf{e_B} \end{cases} \tag{5.10}$$

With respect to the traditional Exergy-based Input-Output analysis introduced in Chap. 4, the Bioeconomic ExIO analysis presents the following peculiarities:

- Since the final demand of the human labour production process is equal to zero, the total primary exergy requirements of the same sector will also result equal to zero. Differently, the specific embodied exergy in one working hour is different than zero and it represents the primary energy-resources requirements caused by the production of one working hour as a *marginal unit*;
- Since a portion of the final demand has been included within the endogenous transactions matrix, specific environmental burdens of all the products will result greater with respect to the results of the standard IO model;
- The equality between the total exergy embodied in goods and services production calculated through the standard and the Bioeconomic models are equal. In other words, application of the Bioeconomic model results in a reallocation of the environmental burdens among final demand products based on their embodied human labour requirements.

References

Ayres, R. U. (2001). The minimum complexity of endogenous growth models: The role of physical resource flows. *Energy, 26*, 817–838.
Ayres, R. U. (2004). On the life cycle metaphor: Where ecology and economics diverge. *Ecological Economics, 48*, 425–438.

Boustead, I., & Hancock, G. F. (1979). *Handbook of industrial energy analysis*. New York: Wiley.

Brown, M. T., & Herendeen, R. A. (1996). Embodied energy analysis and EMERGY analysis: A comparative view. *Ecological Economics, 19*, 219–235.

Consoli, F., Allen, D., Boustead, I., Fava, J., Franklin, W., Jensen, A. A., de Dude, N., Parrish, R., Perriman, R., & Postleithwaite, D. (1993). Guidelines for life-cycle assessment: A 'Code of Practice'. In *Society of environmental toxicology and chemistry (SETAC), Pensacola, FL, USA*.

Costanza, R. (1980). Embodied energy and economic valuation. *Science, 210*, 1219–1224.

Costanza, R., & Herendeen, R. A. (1984). Embodied energy and economic value in the United States economy: 1963, 1967 and 1972. *Resources and Energy, 6*, 129–163.

Dooley, P. C. (2005). *The labour theory of value*.

Duchin, F., & Steenge, A. E. (2007). *Mathematical models in input-output economics*. Troy, NY: Rensselaer Polytechnic Institute.

Faber, M., & Manstetten, R. (2009). *Philosophical basics of ecology and economy*. Routledge.

Fukuda, K. (2003). Production of exergy from labour and energy resources. *Applied Energy, 76*, 435–448.

Georgescu-Roegen, N. (1975). Energy and economic myths. *Southern Economic Journal*, 347–381.

Georgescu-Roegen, N. (1979). Energy Analysis and Economic Valuation. *Southern Economic Journal, 45*, 1023–1058.

Hau, J. L., & Bakshi, B. R. (2004). Promise and problems of emergy analysis. *Ecological Modelling, 178*, 215–225.

Kamp, A., Morandi, F., & Estergård, H. (2016). Development of concepts for human labour accounting in emergy assessment and other environmental sustainability assessment methods. *Ecological Indicators, 60*, 884–892.

King, R. G., Plosser, C. I., & Rebelo, S. T. (1988). Production, growth and business cycles: I. The basic neoclassical model. *Journal of monetary Economics, 21*, 195–232.

Liao, W., Heijungs, R., & Huppes, G. (2012). Thermodynamic analysis of human–environment systems: A review focused on industrial ecology. *Ecological Modelling, 228*, 76–88.

Mayumi, K. (2009). Nicholas Georgescu-Roegen: His bioeconomics approach to development and change. *Development and Change, 40*, 1235–1254.

Miller, R. E., & Blair, P. D. (2009). *Input-output analysis: Foundations and extensions*.

Pokrovskii, V. N. (2011). *Econodynamics: The theory of social production*. Springer.

Pyatt, G., & Round, J. I. (1985). *Social accounting matrices: A basis for planning*.

Rocco, M., Colombo, E., & Sciubba, E. (2014). Advances in exergy analysis: a novel assessment of the extended exergy accounting method. *Applied Energy, 113*, 1405–1420.

Sciubba, E. (2001). Beyond thermoeconomics? The concept of extended exergy accounting and its application to the analysis and design of thermal systems. *Exergy, an International Journal, 1*, 68–84.

Sciubba, E. (2004). From engineering economics to extended exergy accounting: A possible path from monetary to resource-based costing. *Journal of Industrial Ecology, 8*, 19–40.

Sciubba, E. (2011). A revised calculation of the econometric factors α- and β for the extended exergy accounting method. *Ecological Modelling, 222*, 1060–1066.

Sciubba, E., & Ulgiati, S. (2005). Emergy and exergy analyses: Complementary methods or irreducible ideological options? *Energy, 30*, 1953–1988.

Szargut, J., & Morris, D. R. (1987). Cumulative exergy consumption and cumulative degree of perfection of chemical processes. *International Journal of Energy Research, 11*, 245–261.

Szargut, J., Morris, D. R., & Steward, F. R. (1987). *Exergy analysis of thermal, chemical, and metallurgical processes*.

Szargut, J., Ziębik, A., & Stanek, W. (2002). Depletion of the non-renewable natural exergy resources as a measure of the ecological cost. *Energy Conversion and Management, 43*, 1149–1163.

Treloar, G. J. (1998). *A comprehensive embodied energy analysis framework*. Faculty of Science and Technology: Deakin University.

Wales, T. J., & Woodland, A. D. (1977). Estimation of the allocation of time for work, leisure, and housework. *Econometrica: Journal of the Econometric Society*, 115–132.

Xu, M., Williams, E., & Allenby, B. (2009). Assessing environmental impacts embodied in manufacturing and labor input for the China–US trade. *Environmental Science and Technology, 44*, 567–573.

Chapter 6
Case studies: applications of the Exergy based Input-Output analysis

The main objective of this chapter is to perform practical applications of the ExIO framework, highlighting the relevance of Exergy-based Input-Output analysis in Environmental Impact Analysis and Life Cycle Assessment disciplines.

The ExIO framework has been applied to different case studies: (1) analysis of primary exergy embodied in goods and services produced by national economies; (2) applications of Exergy Cost Theory and Exergy Life Cycle Assessment to a Waste to Energy power plant; (3) applications of the Bioeconomic ExIO model to goods and services produced by the Italian economy, and comparative evaluation of alternative options to clean dishes in Italy.

6.1 Exergy-based Input-Output analysis of World national economies

Primary exergy embodied in goods and services produced by different World economies is computed according to Single-Region and Multi-Regional models, as introduced in Sect. 4.1. A focus on economic sectors of Italy is performed, distinguishing among their direct and embodied primary exergy requirements. Results are then compared and discussed.

All the calculations are performed for the reference year 2010, and all the algorithms for the application of ExIO have been developed in *Matlab*®.

6.1.1 Data sources and assumptions

The ExIO framework described in Chap. 4 is here applied for the analysis of different World national economies. In order to allow comparisons among results of different models, productive sectors of the adopted MIOTs must be defined with the

© The Author(s) 2016   101
M.V. Rocco, *Primary Exergy Cost of Goods and Services*,
PoliMI SpringerBriefs, DOI 10.1007/978-3-319-43656-2_6

same aggregation level and compiled according to a unified standard. For such reasons, the *World Input Output Database* (WIOD) is here adopted as the only source of MIOTs (Dietzenbacher and Lahr 2004; Dietzenbacher et al. 2013; Timmer et al. 2015) (http://www.wiod.org/). WIOD is a public and free database: it covers 27 European countries and 13 other major World countries, providing annual data for the period from 1995 to 2009. The database collects data of countries based on data quality and availability, according to their economic importance (they cover more than 80 % of the World GDP).

The WIOD database collects *National Input-Output Tables* (NIOTs) as symmetric industry-by-industry tables in the same format of (Fig. 4.4). NIOTs are compiled in *USD* at current basic price, and they represents each national economy as composed by 35 branches listed in Table 6.1, according to the ISIC rev. 3 standard (which corresponds to the NACE rev. 1) (Eurostat 2008). All data in the NIOTs are obtained from official national statistics and are consistent with the *System of National Accounts* (United Nations, 2009).

Moreover, *World Input-Output Tables* (WIOTs) are also provided by WIOD: these Multi-Regional tables are compiled according to the *multi-directional trades model* (*model e*) and their general structure are visible in Fig. 4.6. Values of gross outputs and intermediate transactions have been properly harmonized across all the countries. Because the analyzed economies do not cover all the countries in the World, a fictitious economy named *Rest of the World* (*RoW*) is added to the WIOT, in order to capture all the World trade flows and to close the balance between total World economic outlays and outputs.

Furthermore, WIOD database collects additional data related to *Socio-economic accounts* and *Environmental accounts*. For the purpose of ExIO analysis, the exergy equivalents of primary fossil fuels endogenous extraction have been considered, namely raw coal, crude oil and natural gas (see Sect. 3.2.3) (Genty et al. 2012). Other ancillary data (TPES, GPD and so on) are taken from *World Bank* and *International Energy Agency* (IEA) databases.

Additional details about the WIOT database can be retrieved in literature (Dietzenbacher et al. 2013; Erumban et al. 2012; Genty et al. 2012; Timmer et al. 2012, 2015). The list of the analyzed countries and the related fundamental data are resumed in Table 6.2: *population*, *Gross Domestic Product* (GDP), *total endogenous fossil fuels production*.

With reference to Table 6.3, exogenous resources matrix $\mathbf{R}_{end,i}$ accounts for primary fossil fuels endogenously produced in year 2010 by the ith national economy, retrieved in the IEA statistics database (IEA 2014) and in WIOD technical documents (Genty et al. 2012). These values are then converted in ktoe of exergy considering their average values of chemical exergy (ex_{ch}) and the reference value of 41,868 MJ/toe (Song et al. 2012). It is assumed that all the primary non-renewable energy-resources are taken from the environment by the ISIC sector *Mining and quarrying* (code B) (Tukker et al. 2006).

In order to understand the results obtained through different international trades models, the fractions of fossil fuels, goods and services traded among the national economies have been calculated according to the following indicators:

Table 6.1 Economic activities considered in NIOTs

No.	NACE code	Name
1	AtB	Agriculture, hunting, forestry and fishing
2	C	Mining and quarrying
3	15t16	Food, beverages and tobacco
4	17t18	Textiles and textile products
5	19	Leather, leather and footwear
6	20	Wood and products of wood and cork
7	21t22	Pulp, paper, paper, printing and publishing
8	23	Coke, refined petroleum and nuclear fuel
9	24	Chemicals and chemical products
10	25	Rubber and plastics
11	26	Other non-metallic mineral
12	27t28	Basic metals and fabricated metal
13	29	Machinery, nec
14	30t33	Electrical and optical equipment
15	34t35	Transport equipment
16	36t37	Manufacturing, nec; recycling
17	E	Electricity, gas and water supply
18	F	Construction
19	50	Sale, maintenance and repair of motor vehicles and motorcycles; retail sale of fuel
20	51	Wholesale trade and commission trade, except of motor vehicles and motorcycles
21	52	Retail trade, except of motor vehicles and motorcycles; repair of household goods
22	H	Hotels and restaurants
23	60	Inland transport
24	61	Water transport
25	62	Air transport
26	63	Other supporting and auxiliary transport activities; activities of travel agencies
27	64	Post and telecommunications
28	J	Financial intermediation
29	70	Real estate activities
30	71t74	Renting of M & Eq and other business activities
31	L	Public admin and defense; compulsory social security
32	M	Education
33	N	Health and social work
34	O	Other community, social and personal services
35	P	Private households with employed persons

Table 6.2 Data of the WIOD countries for the reference year 2010

Countries		Pop (Millions)	GDP (GUSD)	Fossil prod. (Mtoe)	Fossil import (Mtoe)	Fossil export (Mtoe)	$y_{f,imp}$ (%)	$y_{f,exp}$ (%)	$y_{p,imp}$ (%)	$y_{p,exp}$ (%)
Australia	AUS	22.03	797.4	302,553	27,179	226,185	26.2	218.4	9.9	11.3
Austria	AUT	8.39	325.6	2512	20,933	4228	108.9	22.0	23.4	25.3
Belgium	BEL	10.88	400.4	0	59,632	7498	114.4	14.4	30.7	33.2
Bulgaria	BGR	7.40	33.0	5022	9980	47	66.7	0.3	22.8	19.8
Brazil	BRA	195.21	1096.8	124,170	40,135	32,614	30.5	24.8	6.6	6.6
Canada	CAN	34.01	1240.1	328,882	65,616	203,316	34.3	106.3	15.8	15.5
China	CHN	1337.71	3839.3	1,951,023	360,057	49,895	15.9	2.2	8.1	9.8
Cyprus	CYP	0.82	19.2	0	11	0	100.0	0.0	20.9	11.1
Czech Republic	CZE	10.47	148.5	21,201	17,030	5350	51.8	16.3	26.9	29.1
Germany	DEU	81.78	2954.4	60,331	206,615	18,820	83.3	7.6	19.1	24.0
Denmark	DNK	5.55	256.8	19,827	5612	11,208	39.4	78.8	22.9	26.8
Spain	ESP	46.58	1179.2	3466	97,050	2121	98.6	2.2	13.7	11.6
Estonia	EST	1.33	13.9	3943	608	429	14.8	10.4	24.4	27.4
Finland	FIN	5.36	204.2	1846	19,272	6872	135.3	48.2	18.7	20.1
France	FRA	65.02	2204.5	1888	118,666	2787	100.8	2.4	13.5	12.8
United Kingdom	GBR	62.77	2360.0	126,572	118,883	57,780	63.3	30.8	15.4	14.9
Greece	GRC	11.15	241.0	7428	24,699	9592	109.6	42.6	16.9	8.6
Hungary	HUN	10.00	109.3	4917	15,310	483	77.5	2.4	33.5	36.1
Indonesia	IDN	240.68	377.9	309,511	19,991	209,639	16.7	174.9	11.0	12.5
India	IND	1205.62	1243.7	333,418	246,190	1213	42.6	0.2	10.5	9.5
Ireland	IRL	4.56	203.3	1214	8556	112	88.6	1.2	37.4	48.4

(continued)

Table 6.2 (continued)

Countries		Pop (Millions)	GDP (GUSD)	Fossil prod. (Mtoe)	Fossil import (Mtoe)	Fossil export (Mtoe)	$y_{f,imp}$ (%)	$y_{f,exp}$ (%)	$y_{p,imp}$ (%)	$y_{p,exp}$ (%)
Italy	ITA	59.28	1763.9	12,567	161,816	2079	93.9	1.2	13.7	12.8
Japan	JPN	127.45	4648.5	3904	382,764	472	99.1	0.1	6.9	8.1
Korea	KOR	49.41	1098.7	2142	233,799	481	99.3	0.2	18.2	20.5
Lithuania	LTU	3.10	27.5	127	12,188	136	100.1	1.1	28.4	27.3
Luxembourg	LUX	0.51	40.7	0	1263	0	100.0	0.0	51.0	63.3
Latvia	LVA	2.10	15.5	2	1020	3	100.1	0.3	19.7	18.8
Mexico	MEX	117.89	953.1	203,184	17,484	77,711	12.2	54.4	16.4	16.5
Malta	MLT	0.42	6.7	0	0	0	0.0	0.0	33.7	30.0
Netherlands	NLD	16.62	683.1	65,098	94,815	48,678	85.2	43.8	27.3	32.7
Poland	POL	38.18	383.2	59,817	40,865	11,261	45.7	12.6	20.1	20.3
Portugal	PRT	10.64	197.2	0	18,167	0	100.0	0.0	17.0	12.5
Romania	ROU	20.25	114.1	18,706	9076	136	32.8	0.5	17.6	14.5
Russia	RUS	142.39	909.2	1,226,585	19,912	488,345	2.6	64.4	11.0	14.6
Slovak Republic	SVK	5.39	77.1	911	13,701	277	95.6	1.9	28.6	29.6
Slovenia	SVN	2.05	39.0	1202	877	0	42.2	0.0	25.9	26.0
Sweden	SWE	9.38	401.6	238	24,410	875	102.7	3.7	21.2	25.2
Turkey	TUR	72.14	565.1	20,564	61,993	534	75.6	0.7	13.0	9.4
Taiwan	TWN	23.16	377.6	249	99,133	79	99.8	0.1	28.3	32.7
United States	USA	309.33	13,595.6	1,373,187	638,112	83,688	33.1	4.3	8.1	6.2

Table 6.3 ISIC sectors and average values for LHV and ex_{ch} of primary non-renewable energy-resources

Primary fuels	ISIC code	ISIC name	LHV (MJ/kg)	ex_{ch} (MJ/kg)
Raw coal	B 05	Mining of coal and lignite	30.98	34.08
Natural gas	B 0620	Extraction of natural gas	47.82	49.73
Crude oil	B 0610	Extraction of crude petroleum	41.87	44.38

- *Fractions of imported and exported raw fuels* $(y_{f,imp}, y_{f,exp})$: they represent the ratios between the imported and exported primary energy-resources and the net consumption of primary fuels, as defined by relation (6.1). Notice that values of ex_{net} refers to the net primary fossil fuels required by the considered economy, defined as the algebraic sum between endogenous fuels production, imports and exports;

$$ex_{net} = ex_{prod} + ex_{imp} - ex_{exp} \;\rightarrow\; \begin{cases} y_{f,imp} = ex_{imp}/ex_{net} \\[4pt] y_{f,exp} = ex_{exp}/ex_{net} \end{cases} \tag{6.1}$$

- *Fractions of imported and exported products* $(y_{p,imp}, y_{p,exp})$: in a similar fashion, and with reference to relation (6.2), these factors are respectively evaluated as the ratios between imported/exported production and the net total national production of goods and services. Notice that values of p_{net} are here defined as the algebraic sum between endogenous goods and services production, imports and exports;

$$p_{net} = p_{prod} + p_{imp} - p_{exp} \;\rightarrow\; \begin{cases} y_{p,imp} = \dfrac{p_{imp}}{p_{net}} = \dfrac{(\mathbf{Z}\cdot\mathbf{i}+\mathbf{f})_{\mathrm{imp}}^{\mathrm{T}}\mathbf{i}}{\left[(\mathbf{Z}\cdot\mathbf{i}+\mathbf{f})_{\mathrm{end+imp}}-\mathbf{e_N}\right]^{\mathrm{T}}\mathbf{i}} \\[16pt] y_{p,exp} = \dfrac{p_{exp}}{p_{net}} = \dfrac{\mathbf{e_N^{\mathrm{T}}}\mathbf{i}}{\left[(\mathbf{Z}\cdot\mathbf{i}+\mathbf{f})_{\mathrm{end+imp}}-\mathbf{e_N}\right]^{\mathrm{T}}\mathbf{i}} \end{cases} \tag{6.2}$$

Based on numerical values of the above introduced indicators, represented in Fig. 6.1, the following considerations can be made:

- Generally, higher fractions of both imported and exported products ($y_{p,imp}$ and $y_{p,exp}$) correspond to smaller economies. Notably, countries that are more dependent from imported products (high values of $y_{p,imp}$) tend also to export large portions of their net production (high values of $y_{p,exp}$). In general, imported/exported fractions of products are lower than 40 %;

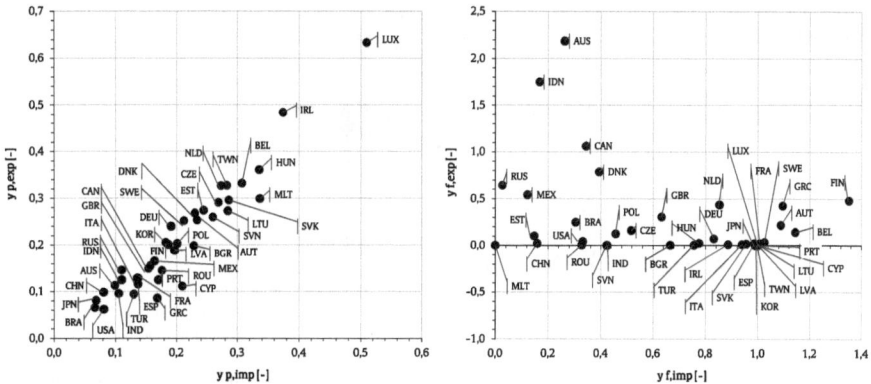

Fig. 6.1 Scatter plots. *Left side* $y_{p,imp}$ versus $y_{p,exp}$. *Right side* $y_{f,imp}$ versus $y_{f,exp}$

- More than half of the considered countries are strongly dependent by the others in terms of primary fossil fuels imports $\left(y_{f,imp} > 50\ \%\right)$. In a different way with respect to the trades of products, countries whose imports primary fossil fuels do not export them to others and vice versa;
- Based on the aforementioned indicators, it can be said that fossil fuels are not the most relevant contributions, in terms of value, with respect to the whole imported economic production. Indeed, many countries with almost 100 % of imported primary fuels result in less than 30 % of the total imported production expressed in value terms (GRC, FIN, SWE, etc.). This issue has been largely debated in literature, and it can be described as an underestimation of the value of primary energy-resources within the economic value of goods and services (Glucina and Mayumi 2010; Mayumi 2009). Apart from discussion about the development of new economic paradigms, this serves here to emphasize the need of developing proper international trade flows treatment methods for the purpose of embodied exergy accounting, since large values of imported fossil energy could be hidden within small economic values of imported products;
- Finally, large errors in the embodied exergy accounting are expected from the application of Single-Region models to countries with high values of imported/exported fractions of products and fossil fuels.

6.1.2   About the evaluation of the Leontief Inverse Coefficients matrix

The evaluation of primary exergy embodied in national goods and services according to the Multi-Regional models here presented requires to derive the Leontief inverse coefficients matrix starting from a very large Input-Output table. It is well established by literature that the application of *direct* matrix inversion

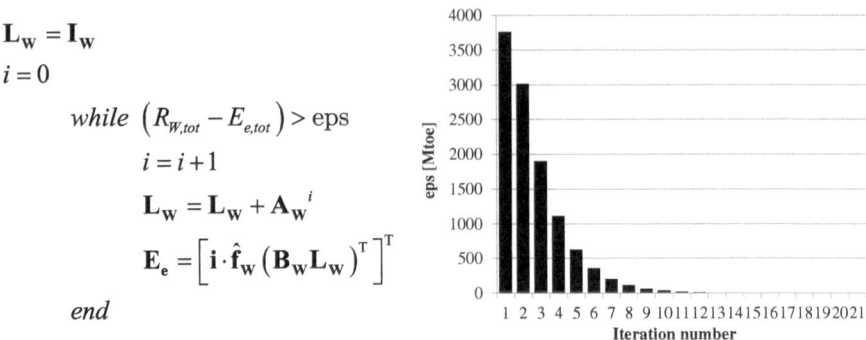

$$L_W = I_W$$

$$i = 0$$

$$while \ \left(R_{W,tot} - E_{e,tot} \right) > eps$$

$$i = i + 1$$

$$L_W = L_W + A_W{}^i$$

$$E_e = \left[i \cdot \hat{f}_W \left(B_W L_W \right)^T \right]^T$$

end

Fig. 6.2 Algorithm adopted for the iterative evaluation of the Leontief Inverse matrix (*left side*) and values of eps resulting after every iteration (*right side*)

methods to such systems are considerably slower and less accurate with respect to *indirect* method (Watkins 2004). Therefore, the evaluation of the Leontief inverse coefficients matrix can be performed here through an iterative technique based on the *Power Series approximation*, represented by relation (6.3) and previously introduced in Sect. 2.3.3. Starting from such expression, the algorithm reported in Fig. 6.2 (left side) have been applied.

$$L_W = (I_W - A_W)^{-1} \rightarrow L_W = \sum_{t=0}^{+\infty} A_W^i = I_W + A_W + A_W A_W + \cdots \qquad (6.3)$$

Convergence criterion for the developed algorithm has been assumed as the difference between the primary exergy produced by all the countries $\left(R_{W,tot} \right)$ and the primary exergy embodied in the World economic production $\left(E_{e,tot} \right)$, evaluated according to relation (4.31). With reference to Fig. 6.2 (right side), such difference results less than 1 Mtoe after the 16th iteration, and convergence is then reached after about 2 min of calculation. Differently with respect to Multi-Regional models, application of Single-Region models requires the evaluation of a Leontief inverse coefficients matrix starting from smaller Input-Output tables, and thus direct inversion methods can be simply adopted.

In depth discussion about efficient algorithms and numerical methods applied to LCA and IOA can be found in the works of *Peters* (Peters 2007) and *Suh and Heijungs* (Heijungs and Suh 2002).

6.1.3 Primary exergy embodied in national economic production

Values of exergy embodied in the whole national economic production have been calculated for all the World countries based on all the international trades models presented in Sect. 4.1. Results are listed in Table 6.4. In order to allow a more meaningful comparison among different countries, embodied exergy of products are quantified in terms of *tons of oil equivalent per capita* (E_{ex}, toe/person/year) rather than per unit of GDP: this because monetary transactions in WIOT are not defined in terms of *purchasing power parity* (PPP), therefore 1 USD worth produced by the same sector of different economies may correspond to different quantities and types of products.

From the theoretical standpoint, results of different international trades models are not comparable, since different assumptions are made and thus different space boundaries are defined according to each model. However, in current LCA practice, results of different models, and even results of Process-based and Input-Output based techniques, are usually compared (Hendrickson et al. 1997, 2010; Lave et al. 2000). Therefore, it is interesting to investigate the differences between numerical results obtained in order to provide a qualitative evidence of the relevance that international trades models may have in evaluating primary exergy embodiment of products in a LCA perspective.

In the following, results of *model e* are assumed as the reference for the evaluation of the accuracy of the other models, since all the hypotheses to treat international trades of products are removed, as can be inferred from Sect. 4.1. With reference to Table 6.4, results of Single-Region *models a* and *b* are very similar to each other, but very different compared to results of *model e*. Indeed, exergy embodied in products of small countries with no endogenous fossil fuels production (e.g. BEL, CYP, LUX, MLT and PRT) is equal to zero, while results for exporting countries are largely overestimated (e.g. AUS, CAN, NLD and RUS). On the other hand, results of Multi-Regional models exhibit a very different behavior: results of *model d* seems to be more accurate with respect to *model c*, which produces generally overestimated results with respect to *model e*.

Relative errors (err_k, in percentage) between results of Single-Region and Multi-Regional models with respect to *model e* have been derived and reported for each kth country in Fig. 6.3, considering ascending values of fossil fuels production $ex_{prod,k}$. Accuracy of all international trades models with respect to *model e* is quantitatively evaluated through the parameters collected in Table 6.5. Beside *maximum*, *minimum* and *average* values of embodied exergy in national production, the *average relative error* $\overline{err}_k$ and the *Pearson's correlation coefficient* r_k are also evaluated. Among Multi-Regional models, *model d* gives better results than *method c*, leading to an average error of about 24 % with respect to *model e*. On the other hand, *model b* results as the best Single-Region model with an average error of 66 %.

Based on the obtained results, the following considerations can be made:

- Relative errors in Fig. 6.3 clearly demonstrate that each model can be better than the others in predicting the average embodied exergy requirements of products, depending on the analyzed country. This depends by a multiplicity of factors: in general, Multi-Regional models return more accurate values than the Single-Region ones for those countries characterized by high fractions of fossil fuels imports (ITA, ESP, IRL, and so on), or with large penetrations of

Table 6.4 Embodied exergy per capita (toe/p/year) evaluated with all the international trades models for the considered World countries. Relative differences of models a, b, c and d with respect to model e are reported in last four columns

Countries	$E_{ex,a}$ (toe/p)	$E_{ex,b}$ (toe/p)	$E_{ex,c}$ (toe/p)	$E_{ex,d}$ (toe/p)	$E_{ex,e}$ (toe/p)	err_{a-e}	err_{b-} e	err_{c-} e	err_{d-} e
AUS	14.36	14.93	16.85	5.76	6.46	122	131	161	−11
AUT	0.31	0.77	3.49	2.02	3.33	−91	−77	5	−39
BEL	0.00	0.00	6.20	2.84	4.91	−100	−100	26	−42
BGR	0.77	1.61	2.74	1.65	1.89	−59	−15	45	−13
BRA	0.66	0.78	0.99	0.66	0.76	−13	3	30	−13
CAN	9.97	11.86	12.83	4.36	5.03	98	136	155	−13
CHN	1.52	2.11	1.93	1.60	1.70	−10	24	14	−6
CYP	0.00	0.00	2.48	2.22	2.89	−100	−100	−14	−23
CZE	2.11	2.96	3.49	1.51	2.33	−9	27	50	−35
DEU	0.99	2.90	2.74	1.77	2.72	−64	7	1	−35
DNK	4.46	4.18	5.70	2.67	4.23	6	−1	35	−37
ESP	0.08	0.53	2.56	1.85	2.46	−97	−78	4	−25
EST	3.44	4.18	4.16	2.74	3.35	3	25	24	−18
FIN	0.37	1.30	7.53	4.23	5.12	−93	−75	47	−17
FRA	0.03	0.14	2.25	1.75	2.55	−99	−94	−12	−32
GBR	2.20	3.24	3.97	2.21	2.80	−22	16	42	−21
GRC	0.77	3.16	3.41	2.86	3.32	−77	−5	3	−14
HUN	0.58	3.87	3.36	2.23	2.81	−79	38	20	−21
IDN	1.31	1.37	1.44	0.68	0.75	75	83	93	−9
IND	0.29	0.49	0.38	0.29	0.32	−10	56	21	−7
IRL	0.28	0.60	2.84	1.74	2.80	−90	−79	1	−38
ITA	0.23	1.82	5.05	4.04	4.60	−95	−61	10	−12
JPN	0.03	0.17	3.63	2.99	3.39	−99	−95	7	−12
KOR	0.05	1.24	5.58	3.21	3.84	−99	−68	45	−16
LTU	0.05	0.72	6.11	3.28	3.64	−99	−80	68	−10
LUX	0.00	0.00	1.77	0.69	2.71	−100	−100	−35	−75
LVA	0.00	0.00	1.10	0.85	1.51	−100	−100	−27	−44
MEX	1.89	1.76	2.01	1.14	1.34	41	31	49	−15
MLT	0.00	0.00	0.58	0.39	1.98	−100	−100	−71	−80
NLD	4.00	9.22	12.43	4.73	5.86	−32	57	112	−19
POL	1.68	2.68	3.22	2.06	2.40	−30	12	34	−14
PRT	0.00	0.00	1.28	1.00	1.48	−100	−100	−14	−33
ROU	1.03	1.50	1.64	1.26	1.55	−34	−3	5	−19
RUS	8.97	8.68	8.70	2.15	2.21	306	293	294	−3
SVK	0.19	0.84	3.59	2.18	2.80	−93	−70	28	−22
SVN	0.67	1.17	1.93	1.32	2.29	−71	−49	−15	−42

(continued)

Table 6.4 (continued)

Countries	$E_{ex,a}$ (toe/p)	$E_{ex,b}$ (toe/p)	$E_{ex,c}$ (toe/p)	$E_{ex,d}$ (toe/p)	$E_{ex,e}$ (toe/p)	err_{a-e}	err_{b-e}	err_{c-e}	err_{d-e}
SWE	0.03	0.05	4.33	2.13	3.35	−99	−98	29	−36
TUR	0.31	0.30	0.52	0.39	0.56	−45	−45	−6	−30
TWN	0.01	0.08	4.89	2.21	2.82	−100	−97	73	−21
USA	4.66	7.16	6.77	5.94	6.38	−27	12	6	−7

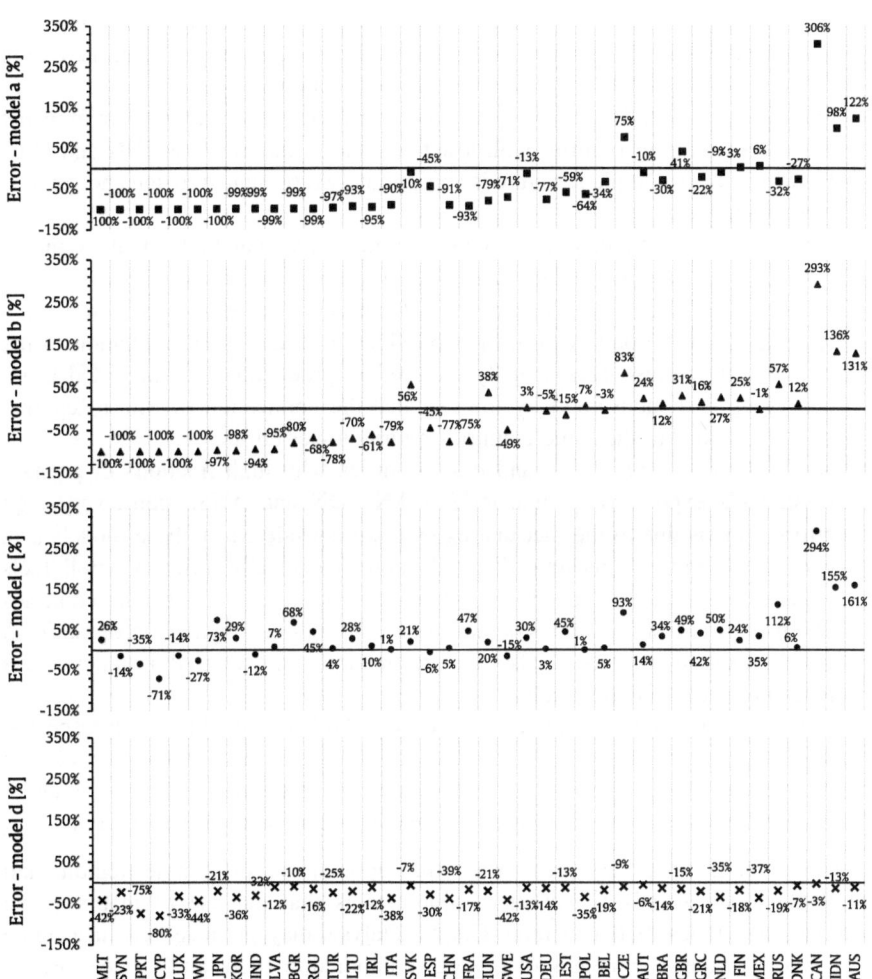

Fig. 6.3 Relative errors in embodied exergy per capita of models (**a–d**) with respect to model (**e**). Results are displayed with respect to ascending values of fossil fuels production ex_{prod}

Table 6.5 Comparison of results obtained with all the international trades models

Indicator	Units	a	b	c	d	e
Max value	toe/p	14.36	14.93	16.85	5.94	**6.46**
Min value	toe/p	0.00	0.00	0.38	0.29	**0.32**
Average embodied exergy per capita	toe/p	1.71	2.46	4.16	2.24	**2.93**
Average relative error	%	75 %	66 %	43 %	24 %	–
Correlation coefficient (r)	–	0.477	0.586	0.821	0.944	–

alternative power sources (FRA, which is strongly based on nuclear energy). Moreover, the relative error of Single-Region models is generally an increasing function of the endogenous fossil energy production ex_{prod};

- On average, Single-Region *model a* results in larger errors with respect to *model b*, except for large countries with low fractions of imports, like CHN and IND, for which the economic structure is appropriately represented by the endogenous Input-Output table;
- On average, the application of Single-Region *models a* and *b* to small countries with high values of non-renewable energy imports $y_{f,imp}$ cause underestimations of the embodied exergy of production. This because the endogenous productive structure of such countries does not provide an adequate representation of the primary energy-resources actually required in producing their imported products. This is demonstrated by the fact that the application of Multi-Regional *models c* and *d* to the same countries returns lower errors;
- Application of *models a*, *b* and *c* to countries with high fractions of primary fossil fuels exports $y_{f,exp}$, like RUS, CAN, IDN and AUS, return very high errors. This is due to the fact that exports are considered by these models as a part of the endogenous final demand: more realistically, primary fossil fuels exports are devoted to sustain the intermediate requirements of other countries (mainly for industry). This is demonstrated through the application of *model d*, which takes into account also the destinations of exported products, thus returning very accurate results;
- *Model d* slightly underestimates the exergy embodied in products with respect to *model e* for all the considered countries: this is due to the fact that the exported products exchanged among all the economies different than the one under investigation are considered as part of the endogenous final demand only.

Concluding, what emerges from the application of Single-Region and Multi-Regional models to national economies is that the use of inappropriate international trades model in Input-Output analysis may produce very inaccurate estimations of the specific and total exergy embodied in goods and services. Based on the obtained results, the following general guidelines can be identified: Single-Region *model b* returns acceptable results if applied for the analysis of large economies with low values of imported/exported fuels and products. Considering Single-Region models, *model b* is generally preferable with respect to *model a*. On the other hand, considering Multi-Regional models, the choice among *model c* and

d depends on the objectives of the analysis: in case of LCA applications *model c* and *d* returns similar results, while in case of analysis of trades in embodied exergy among national economies *model d* is preferable with respect to *model c*.

In general, Multi-Regional models are not always more accurate than Single-Region ones, and the latter could be employed for many economies without returning relevant errors in the results.

6.1.4 Exergy embodied in detailed products of the Italian economy

Accounting for the amount of exergy embodied in detailed products of national economies is one of the purposes of LCA analysis. Specific and total exergy embodied in detailed products of the Italian economy in 2010 have been calculated with all the international trades models introduced in Sect. 4.1: results are graphically presented in Fig. 6.4, and reported in Table 6.6. Specific embodied exergy is quantified in *kg of oil equivalent per 100 USD of product*, while the total embodied exergy is quantified in *kton of oil equivalent*.

Based on the obtained results, the following comments can be made. Numerical values of *specific* embodied exergy resulting from the application of Multi-Regional *models c* and *d* are lower with respect to results of *model e*. On the other hand, the application of the same models returns both overestimated and underestimated numerical values of *total* embodied exergy of production. This happens because the final demand vector in *model e* is characterized by the lowest values with respect to the other Multi-Regional models, which in turn classify as final demand large portions of exported national products actually used as intermediate consumptions.

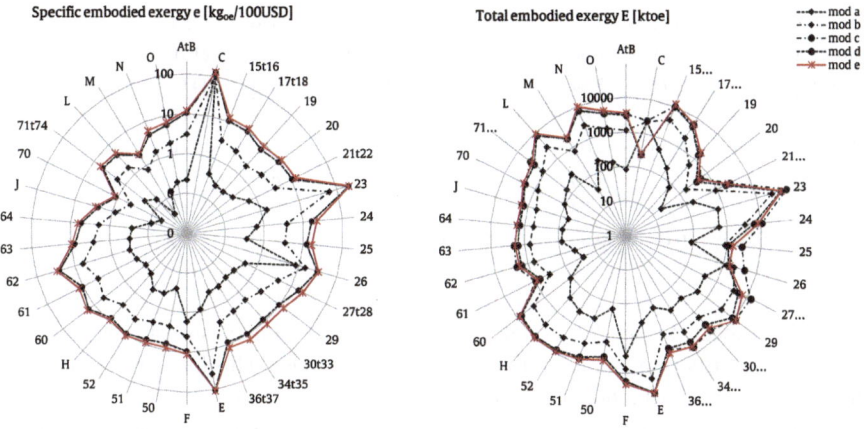

Fig. 6.4 *Left side* specific embodied energy of Italian economic products, expressed in $kg_{oe}/100$ USD

Table 6.6 Specific and total embodied exergy in Italian products in 2010

Code	e (kgoe/100USD)					E (ktoe)				
	Mod a	Mod b	Mod c	Mod d	Mod e	Mod a	Mod b	Mod c	Mod d	Mod e
AtB	0.22	3.13	10.65	10.67	12.27	82	1163	3363	3078	3539
C	116.88	113.45	128.20	128.22	129.57	2347	2278	2437	252	254
15t16	0.40	3.19	10.61	10.65	12.61	497	3954	10,487	10,248	12,133
17t18	0.25	2.72	9.94	9.98	12.18	197	2122	6113	5279	6443
19	0.25	1.90	7.01	7.05	9.01	78	586	1721	1465	1872
20	0.32	2.69	9.24	9.29	11.56	20	169	566	403	501
21t22	0.64	3.59	10.75	10.79	13.00	161	899	2579	1914	2305
23	1.43	64.27	206.90	206.97	210.62	663	29,770	84,054	57,353	58,364
24	0.67	3.88	16.72	16.82	22.67	567	3273	10,748	6108	8237
25	0.37	3.49	12.20	12.28	16.71	89	846	2728	1083	1473
26	8.25	14.48	31.00	31.05	33.49	1196	2100	4316	1394	1504
27t28	0.49	4.34	16.27	16.35	21.51	407	3576	13,044	5170	6799
29	0.35	2.59	9.60	9.67	13.24	464	3463	11,206	8728	11,958
30t33	0.31	1.81	8.65	8.71	12.29	279	1607	5102	3083	4354
34t35	0.36	2.01	9.51	9.58	13.67	348	1920	6344	4240	6046
36t37	0.37	2.47	8.68	8.74	11.95	177	1166	3508	3192	4365
E	0.87	38.78	104.37	104.41	106.79	368	16,321	43,183	42,714	43,688
F	1.66	3.97	9.22	9.25	11.12	3060	7291	16,900	16,871	20,270
50	0.25	2.36	6.60	6.63	8.41	145	1359	3790	3775	4788
51	0.41	2.46	7.32	7.35	8.78	376	2263	6552	5981	7144
52	0.44	3.46	9.26	9.28	10.52	395	3070	8186	7935	8995
H	0.23	2.72	7.76	7.78	8.83	275	3259	9220	9241	10,488
60	0.21	4.78	14.58	14.61	16.54	110	2553	7700	7268	8231
61	0.27	3.51	10.19	10.23	12.46	25	322	929	672	819
62	0.34	6.05	25.50	25.55	29.23	37	653	2249	1562	1786
63	0.30	2.64	7.83	7.86	9.43	80	700	2032	1577	1892
64	0.26	1.87	5.43	5.45	6.56	77	548	1546	1415	1702
J	0.09	0.81	2.37	2.37	2.77	59	546	1559	1391	1623
70	0.05	0.41	1.08	1.09	1.21	105	808	2131	2126	2365
71t74	0.22	1.74	5.14	5.16	6.15	156	1214	3479	2480	2957
L	0.14	1.71	4.66	4.67	5.40	248	2982	8147	8166	9438
M	0.04	0.76	2.03	2.03	2.16	37	745	1991	1991	2113
N	0.13	1.60	4.50	4.52	5.79	218	2683	7566	7600	9730
O	0.20	2.03	5.78	5.80	6.86	143	1436	4022	3959	4682
P	0.00	0.00	0.00	0.00	0.00	0	0	0	0	0

With respect to *model e*, all the other models underestimate the *specific* embodied exergy of national products (Fig. 6.4, left side). This happens because the final demand vector in *model e* is characterized by lowest values with respect to the other models, which in turn classify at portions of exported national products actually directed to intermediate consumption as final demand.

Values of *specific* embodied exergy resulting from the application of Multi-Regional *models c* and *d* are almost equal for all the sectors of the economy: this reveals that specific embodied exergy of products actually depends only by the structure of the endogenous economy and by the sources of imported products, but it is independent from the purposes and the directions of exports. Therefore, it can be said that for the purposes of LCA, accuracy of results of Multi-Regional *model c* is the same of *model d*. On the other hand, if the purpose of the analyst is to account for trades in embodied exergy among national economies, *model d* returns more accurate results with respect to *model c*.

For all the international trades models, highest values of specific embodied exergy result for sectors that are closer to the environment (*primary* sectors), such as sectors C (*Mining and Quarrying*), 23 (*Coke, Refined Petroleum and Nuclear Fuel*) and F (*Electricity, Gas and Water Supply*). Notice that values of total embodied exergy for sector C (*Mining and Quarrying*) calculated according to *models d* and *e* are very different with respect to the other models: this happens because products exported by sector C are primarily used to feed intermediate requirements of other economies, rather than for final uses.

With reference to Sect. 2.3.3, values of specific and total embodied exergy in products can be decomposed by considering the power series expansion of relation (2.34): direct and indirect exergy requirements due to the production of any products of the economy can be thus calculated through relations (6.4) and (6.5) (see Table 6.7).

$$\mathbf{e} = \mathbf{e}_{direct} + \mathbf{e}_{indirect} = \mathbf{B}(\mathbf{I} + \mathbf{A}) + \mathbf{B}(\mathbf{A} + \mathbf{AA} + \mathbf{AAA} + \cdots) \qquad (6.4)$$

$$\mathbf{E} = \mathbf{E}_{direct} + \mathbf{E}_{indirect} = \mathbf{B}(\mathbf{I} + \mathbf{A})\mathbf{f} + \mathbf{B}(\mathbf{A} + \mathbf{AA} + \mathbf{AAA} + \cdots)\mathbf{f} \qquad (6.5)$$

Direct and Indirect primary exergy requirements of Italian products in 2010 have been calculated according to Multi-Regional *model e*: results are graphically presented in Fig. 6.5, and numerical results are reported in Table 6.7. Based on the obtained results, it can be inferred that primary exergy is directly invoked for final uses only by few sectors of the economy, namely sectors C, 23, 26 and E. Primary exergy requirements of other sectors of the economy are caused indirectly, through intermediate consumptions of other goods and services. This serves here to emphasize that *all* the sectors of the economy may contribute to the reduction of primary exergy requirements of the national economy: beside traditional engineering optimization of energy conversion systems, also the implementation of appropriate policies and the development of innovative ways to produce goods and services could play a significant role. Indeed, values of embodied exergy of goods and services reveals where primary energy-resources are actually required to sustain national production: in the perspective of fossil energy-resources displacement, a proper and accurate evaluation of embodied exergy is thus crucial for the purpose of rationalizing and reducing primary exergy requirements of national economies.

Table 6.7 Direct and Indirect contributions to specific and total embodied exergy of Italian products in 2010

Code	e_{direct} (kgoe/100USD)	$e_{indirect}$ (kgoe/100USD)	e (kgoe/100USD)	E_{direct} (ktoe)	$E_{indirect}$ (ktoe)	E (ktoe)
AtB	0.05	12.22	12.27	15	3525	3539
C	118.10	11.46	129.57	232	22	254
15t16	0.10	12.51	12.61	100	12,033	12,133
17t18	0.07	12.11	12.18	35	6408	6443
19	0.05	8.96	9.01	11	1861	1872
20	0.08	11.48	11.56	3	498	501
21t22	0.53	12.47	13.00	94	2211	2305
23	168.43	42.19	210.62	46,673	11,691	58,364
24	1.26	21.41	22.67	458	7778	8237
25	0.15	16.56	16.71	13	1460	1473
26	10.96	22.53	33.49	492	1012	1504
27t28	3.45	18.06	21.51	1091	5708	6799
29	0.18	13.06	13.24	161	11,797	11,958
30t33	0.16	12.13	12.29	56	4298	4354
34t35	0.14	13.53	13.67	60	5986	6046
36t37	0.27	11.68	11.95	99	4266	4365
E	75.57	31.22	106.79	30,915	12,773	43,688
F	1.14	9.98	11.12	2081	18,189	20,270
50	0.08	8.33	8.41	45	4743	4788
51	0.26	8.51	8.78	214	6931	7144
52	0.36	10.17	10.52	304	8691	8995
H	0.04	8.80	8.83	46	10,442	10,488
60	0.04	16.50	16.54	22	8209	8231
61	0.04	12.43	12.46	3	816	819
62	0.08	29.15	29.23	5	1781	1786
63	0.08	9.35	9.43	16	1875	1892
64	0.06	6.50	6.56	15	1687	1702
J	0.02	2.75	2.77	10	1613	1623
70	0.01	1.19	1.21	28	2338	2365
71t74	0.09	6.07	6.15	42	2915	2957
L	0.03	5.37	5.40	50	9387	9438
M	0.01	2.15	2.16	10	2103	2113
N	0.03	5.75	5.79	56	9674	9730
O	0.06	6.80	6.86	42	4640	4682
P	0.00	0.00	0.00	0	0	0

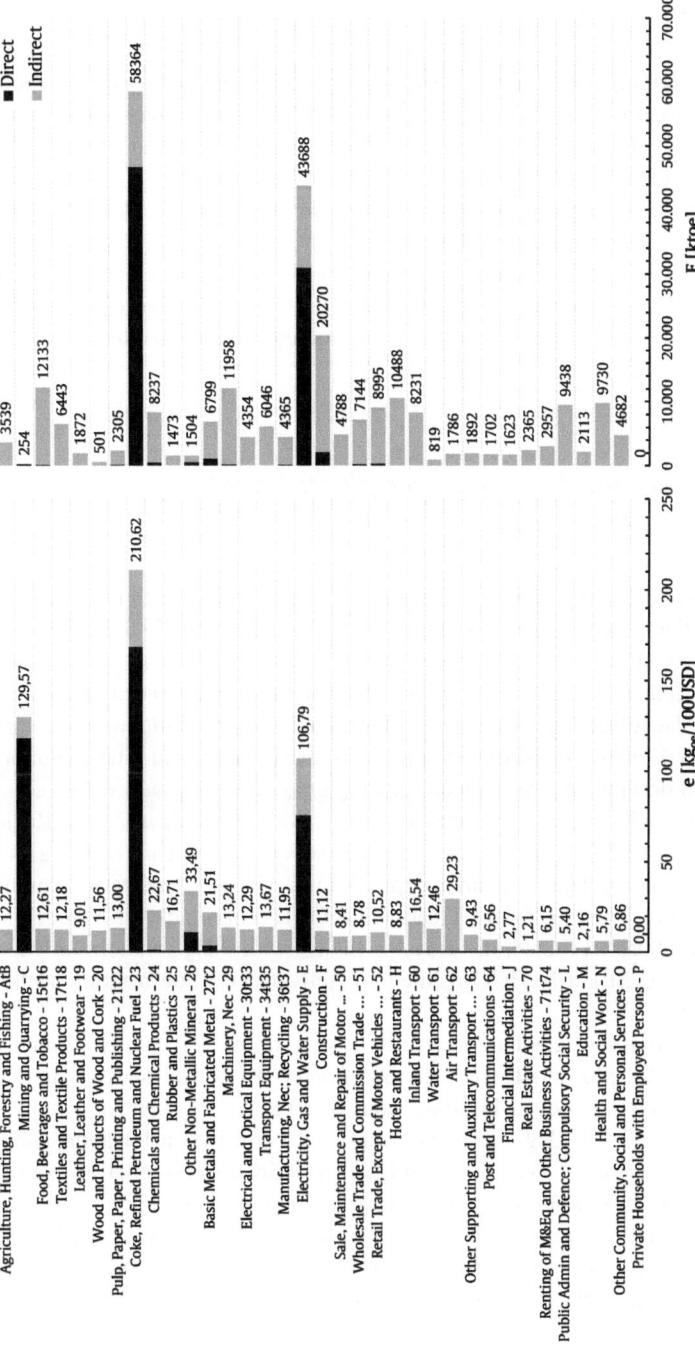

Fig. 6.5 Direct and Indirect contributions to specific and total embodied exergy of Italian products in 2010

6.2   Exergy Life Cycle Assessment of a Waste-to-Energy power plant

In this section, direct and primary exergy costs of electricity production of a Waste-to-Energy power plant are evaluated through conventional Exergy Cost Theory and Exergy LCA analysis. The latter is applied through a Hybrid Input-Output model, as described in Sects. 4.2 and 4.3.2.

Starting from a given design configuration of the power plant, thermodynamic model has been defined and economic cost assessment has been performed. Subsequently, both direct and primary exergy embodied in electricity produced by the system have been evaluated. Conclusively, initial design configuration of the plant has been iteratively improved, verifying each step according to the indicators introduced in Sect. 4.3.2.

6.2.1   Waste-to-Energy power plant: layout and thermodynamic model

The Waste-to-Energy (WtE) plant here described and analyzed is actually operating in the Italian context: therefore, many detailed data related to the plant design have been retrieved from Italian technical literature and from plant operators. In Fig. 6.6 the most relevant components of the WtE plant are depicted. Plant specifications given in the following are related to the initial design configuration of the plant.

The WtE plant is endowed with two waste treatment lines, comprising an air-cooled downward reverse-reciprocating grate (Martin grate) with a counter-flow combustion chamber. The furnace works with an Exhaust Gas Recirculation (EGR) ratio of 15 %. Both lines produce superheated steam sunning in a single Rankine steam cycle: the steam is expanded in a 10 MW net electric nominal power steam turbine. The Flue Gases Treatment (FGT) section includes a Dry Process, and it includes Selective Catalytic Reactor (SCR) with NH_3-solution, an Electrostatic Precipitator (ESP), a $NaHCO_3$/Lime and Activated Carbon reacting section, and a Fabric Filter (FF).

The plant is designed to treat 120 kt/year of waste, mainly *Municipal Solid Waste* (MSW), plus small fractions of *Clinical Waste* and *Sewage Sludge*, which result in an average Lower Heating Value of 10.8 MJ/kg; chemical exergy has been estimated as 12.9 MJ/kg according to a unified correlation for approximate determination of chemical exergy of solid fuels provided by Song et al. (2012).

The entire process allows a waste mass reduction of 94–95 % before landfilling. The WtE plant has a nominal capacity of about 15 t/h of waste, and an availability factor of 0.91. The steam generator produces super-heated steam at 390 °C and 40 bar. Steam exits the turbine at the conditions of 53.5 °C and 0.147 bar. Bleedings at intermediate pressures are designed to provide the necessary heat input to the combustion air pre-heaters, the sludge indirect dryer, and the regenerator.

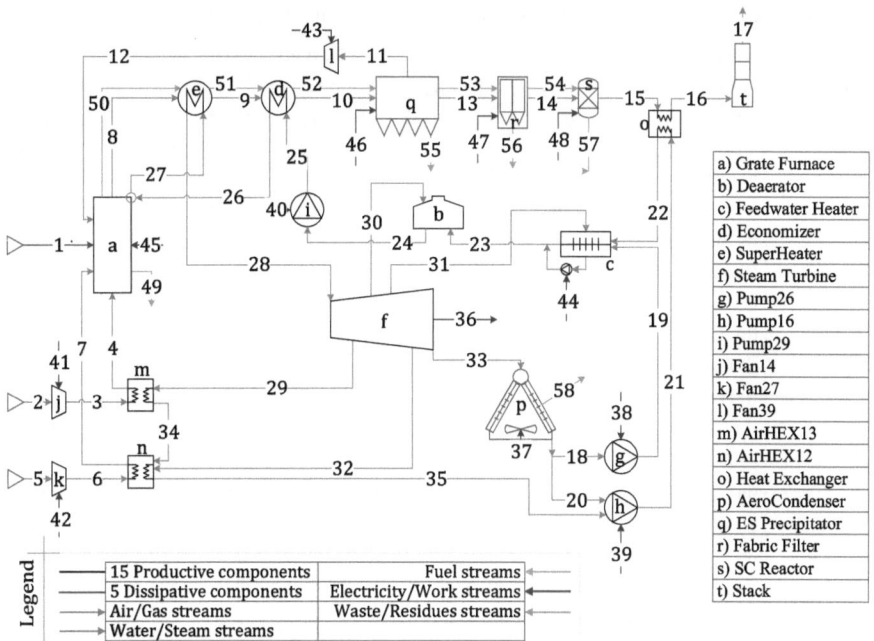

Fig. 6.6 Physical layout of the analyzed waste-to-energy power plant

Reference temperature and pressure for exergy analysis purposes are respectively equal to 25 °C and 1 atm. Chemical composition of the reference environment has been assumed based on appropriate literature (Szargut et al. 2005a, b).

Thermodynamic model of the system has been developed through *ThermoFlow ThermoFlex*® software. A more detailed description of Tecnoborgo WtE plant and its thermodynamic model is not pertinent with the scope of this book and can be retrieved in literature (Buonanno et al. 2009; Tecnoborgo 2013).

6.2.2   Application of the Exergy Cost Theory

Systems of equations for the application of the Exergy Cost Theory have been derived according to the state-of-the-art guidelines provided by literature (Coronado et al. 2013; Kostowski et al. 2014; Torres et al. 2013; Usón and Valero 2011; Valero et al. 2013).

All the material and energy interactions have been classified according to the RPL criterion as introduced in Sect. 4.3.1 (see Table 6.8). The Input-Output table of the system can be compiled and Leontief model can be applied to account for the direct exergy requirements of the plant that are embodied in each product of its components (i.e. the exergy costs of products). Notice that exogenous resources

Table 6.8 RPL classification of the exergy interactions within the system

	Component	Resource	Ex (kW)	Product/loss	Ex (kW)
Productive	(a) Grate furnace	$1 + 4 + 7 + 12 + 45$	55540.3	$8 + (27 - 26) + 49$	28,196.3
	(b) Deareator	30	668.3	24–23	607.9
	(c) Feedwater heater	$31 + 44$	352.1	$23–(22 + 19)$	300.5
	(d) Economizer	9–10	3363.4	26–25	2827.4
	(e) Super heater	8–9	3871.2	28–27	3238.8
	(f) Steam turbine	$28–(29 + \cdots + 33)$	13810.3	$36 + \cdots + 48$	10811.1
	(g) Pump26	38	4.4	19–18	2.6
	(h) Pump16	39	4.8	$2–(20 + 35)$	2.8
	(i) Pump29	40	115.5	25–24	77.7
	(j) Fan14	41	66.7	3–2	49.5
	(k) Fan27	42	44.8	6–5	323.0
	(l) Fan39	43	9.4	12–11	7.2
	(m) AirHEX13	29–34	682.7	4–3	332.2
	(n) AirHEX12	$32 + 34–35$	156.1	7–6	76.8
	(o) Heat exchanger	15–16	165.0	22–21	78.2
Dissipative	(p) Aerocondenser	$33–(18 + 20) + 37$	4832.4	58	4710.0
	(q) ES precipitator	$10–(11 + 13) + 46 + 52–53$	338.8	55	250.0
	(r) Fabric filter	$13–14 + 47 + 53–54$	72.7	56	0.8
	(s) SC reactor	$14–15 + 48 + 54$	53.3	57	0.2
	(t) Stack	16	5871.2	17	5865.7

input to the plant only consists in the exergy content of the waste, while useful products (final demand) consist in the exergy of net electric energy and bottom ashes. All the other outputs are classified as dissipations, such as rejected heat, dusts, fly-ashes and chemical residues.

For each component, specific and total exergy costs $\left(e_S, \dot{E}_S\right)$, exergy destructions $\left(\dot{E}x_{S,(D+L)}\right)$, and exergy cost of exergy destructions $\left(\dot{E}_{S,(D+L)}\right)$ have been computed and reported in Table 6.9. Based on these results, the following considerations can be made:

- Total exergy costs are allocated only to those products that are part of the final demand of the system (i.e. bottom ashes and electricity). Moreover, total exergy cost of products exiting dissipative components results equal to zero, as a consequence of the reallocation procedure: exergy costs of losses have been reallocated to productive components that actually feed the dissipative components, according to a proportionality criterion (Torres Cuadra et al. 2008). Notice that specific exergy costs of dissipative components have no practical meaning and are hence not reported in Table 6.9.

Table 6.9 Results of the application of the ECT to the waste-to-energy plant

Productive components	e_s (J/J)	E_s (kW)	$Ex_{S,(D+L)}$ (kW)	$E_{S,(D+L)}$ (kW)
(a) Grate furnace	3.3	1587.8	27093.1	90354.2
(b) Deareator	5.2	–	60.4	315.9
(c) Feedwater heater	5.5	–	51.7	284.9
(d) Economizer	4.9	–	536.0	2616.2
(e) Super heater	4.9	–	632.4	3099.0
(f) Steam turbine	5.1	52288.8	2999.2	15274.2
(g) Pump26	9.7	–	1.8	17.8
(h) Pump16	9.6	–	0.0	0.0
(i) Pump29	8.5	–	37.8	320.6
(j) Fan14	6.9	–	17.2	118.4
(k) Fan27	6.9	–	11.8	81.8
(l) Fan39	6.7	–	2.3	15.0
(m) AirHEX13	8.1	–	350.5	2828.3
(n) AirHEX12	8.0	–	79.4	633.9
(o) Heat exchanger	8.0	–	86.8	690.0

- Exergy destructions and losses result highest for grate furnace, steam turbine and super heater: these components cause about 96 % of the total thermodynamic irreversibilities within the plant. Therefore, any attempt to improve the plant thermodynamic efficiency should focuses on such components. Moreover, values of the exergy cost of exergy destructions and losses for each component confirm the relevance of grate furnace and steam turbine.
- Economizer and underfire air preheater account for about 3 % of total irreversibilities. While exergy destructions of the economizer are greater with respect to the air preheater, values of the exergy costs of exergy destructions are oppositely ranked. This means that improving efficiency of air preheater rather than the economizer may result in greater reduction of the overall plant irreversibilities. The same opposite ranking between exergy destructions and exergy costs of exergy destructions happens with other components, such as Deareator, Feedwater Heater and Pump29.

6.2.3 Application of the Exergy Life Cycle Assessment

ELCA method is here applied to the Waste-to-Energy power plant according to the Hybrid Input-Output analysis described in Sect. 4.2. The following fundamental hypotheses are assumed for the analysis:

- The whole construction phase of the WtE plant takes place overnight, in the year 2010;
- The prospected operative lifetime of the plant is 25 years at nominal operative conditions;
- Due to a lack of reliable economic data and to the related large uncertainties, disposal phase of the plant life-cycle has been neglected. Thus, primary exergy cost of electricity has been evaluated according to a *cradle-to-gate* approach;

Setup of the Hybrid Input-Output table

The Hybrid Input-Output table should be defined for every life cycle phase of the WtE plant, resembling the general outline reported in Fig. 4.7. In this case, the Hybrid Input-Output table is defined for the construction and the operation phases of the WtE plant. Main data and assumptions required to define the Hybrid system are defined in the following:

- *National input-output table* Z_N, f_N, x_N. The symmetric Italian MIOT of 2010 is taken from the Italian official statistics bureau and it is adopted as the reference to model the Italian supply chains. Italian economy is schematized with 63 productive sectors, classified according to ISIC Rev.4 standard. Transactions of goods and services are expressed in M€ at basic prices according to 2010 exchange rates, including imports from other countries. Since the WtE plant is currently operating within the national economy, this table should be properly adjusted to avoid double counting errors, as described in Sect. 4.2.2. According to the ISIC Rev.4 standard, construction phase of the plant is classified within the economic sector *Construction - Civil engineering* (code F-42), which includes the construction of power plants and waste-treatment facilities. On the other hand, operation phase is classified within the sector *Electricity, gas, steam and air conditioning supply* (code D-35). The Italian MIOT of 2010 is adopted as a reference for both the construction and the average operation phases;
- *National exogenous transactions vector* R_N. It accounts for the primary non-renewable resources absorbed by the national economy. In this case study, international trades of products are treated according to the Single-Region *model b* (see Sect. 4.1.2), so the national exogenous transactions matrix R_N accounts only for the primary fossil fuel endogenously produced in year 2010 by the Italian economy, properly converted in ktoe of exergy. Such values are equal for both construction and operation phases. Data have been retrieved from IEA statistics database (IEA 2010) and reported in Table 6.10. Average values of LHV and Chemical exergy of fuels have been taken from literature (Song et al. 2012). According to ISIC Rev. 4 standard, primary non-renewable resources enter national economy through sector *Mining and quarrying* (code B);
- *System input-output table* Z_S, f_S, x_S. For the construction phase, endogenous transactions matrix Z_S is composed by just one scalar equal to zero, while vectors f_S and x_S are scalars equal to 1, as the useful product of the construction phase only consists in one unit of WtE power plant. Since the production of such plants is not actually part of the Italian final demand, national final demand

Table 6.10 Properties, ISIC sectors and values of primary non-renewable resources produced by the Italian economy in 2010

Primary fuel	ISIC code	ISIC name	LHV (MJ/kg)	ex_{ch} (MJ/kg)	$R_{N,i}$ (ktoe)
Raw coal	B 05	Mining of coal and lignite	30.98	34.08	69.1
Natural gas	B 0620	Extraction of natural gas	47.82	49.73	5957.2
Crude oil	B 0610	Extraction of crude petroleum	41.87	44.38	7158.3
Total	**B**	**Mining and quarrying**	–	–	**13184.6**

vector f_N does not need any adjustment. For the operation phase, the exergy production balance is defined on yearly basis by means of exergy, and it is assumed to be constant for all the 25 estimated operative years. The system endogenous transactions Z_S are equal to zero, since the WtE plant does not consume its own product. The final demand of the system f_S in the operation phase includes only electric energy generation: to avoid double counting errors, national final demand vector f_N must then be adjusted by taking the net electric energy production of the WtE plant, properly converted in monetary units using an average electricity price of 190 €/MWh (this value has been taken from Italian official statistics department), and subtracting this value from the national final demand for electricity according to relation (4.37);

- *System exogenous transactions vector* R_S. It is defined in the same way for both construction and operation phases, assuming that the system does not absorb non-renewable resources directly from the environment. Differently from the Exergy Cost Theory, R_S thus results a scalar equal to zero for both construction and operation phases. This implies that the national exogenous transactions vector does not need to be adjusted according to relation (4.39);
- *Upstream Cutoff matrix* C_{NS}. It represents the transactions of goods and services invoked from detailed sectors of Italian economy by the WtE system during construction and operation phases. Such values are expressed by means of their monetary equivalents, derived through a detailed economic analysis of the WtE plant described in the following paragraph;
- *Downstream Cutoff matrix* C_{SN}. It includes the transactions of goods and services, expressed by means of exergy, that are produced by the WtE plant and absorbed by the Italian economy as intermediate transactions. For the construction phase, C_{SN} is an empty vector, while during the operation phase it includes the exergy content of the bottom ashes recovered from the grate furnace (stream 49). Bottom ashes are considered as by-products recirculated into Italian economy as building material, feeding the ISIC sector *Construction—Civil engineering* (code F 42): it is not necessary to include this term in the hybrid final demand vector, given that it is not the final product of main interest.

Focus on the Upstream Cutoff matrices: economic analysis of the WtE plant
A detailed economic analysis of the WtE plant has been performed to define

Upstream cutoff matrices C_{NS} for construction and operation phases. according to literature guidelines (Green and Perry 1999; Peters et al. 2003).

For the construction phase, the *Total Capital Investment* (TCI) has been evaluated, including purchase, delivery, and installation of components, piping, buildings, land acquisition and yard improvements, electrical systems, instrumentation and control, plus other non-manufacturing costs (e.g. engineering, legal expenses, contingencies, etc.). For the operation phase, the *Total yearly Production Cost* (TPC) has been evaluated, comprising raw materials, labour, utilities, maintenance, depreciation, taxes, insurances, and so on. Values of TCI and TPC result respectively 145.09 M€ and 9.94 M€/y, considering the 2010 exchange rate: based on such values, the total economic cost of electricity results 101.40 €/MWh. Considering the 2010 Italian policy based on Green Certificates and a constant Gate Fee of 100 €/ton of treated MSW (Consonni 2012), total annual revenues are about 22.18 M€/year and the plant Pay Back Time is about 13 years. Results of the economic analysis are then in line with average values retrieved in literature (Bohm et al. 2010; Massarutto et al. 2011; Tsilemou and Panagiotakopoulos 2006).

For the purpose of ELCA, Upstream Cutoff matrix for construction and operation phases has been defined using values listed in Tables 6.11 and 6.12. Monetary transactions in the Italian MIOT are expressed in basic prices: to be consistent, Upstream Cutoff matrices does not include monetary costs due to taxes (mainly VAT), Gate Fee, labor, financing and depreciation.

Table 6.11 Cost categories of TCI included in the upstream cutoff matrix for the construction phase (k€ of 2010)

ISIC Rev.4 reference sector—Code		Monetary cost category	Value (k€)
Manufacture of machinery and equipment n.e.c.	C 28	Purchase equipment cost	49,535
Repair and installation of machinery and equipment	C 33	Purchased equipment installation	10,402
Manufacture of electrical equipment	C 27	Instrumentation and control	11,111
Manufacture of electrical equipment	C 27	Electrical Systems	7802
Manufacture of fabricated metal products	C 24	Piping	9908
Architectural and engineering activities ...	M 71	Engineering and supervision	10,402
Legal and accounting activities	M 69	Legal expenses	718
Electricity, gas, steam and air conditioning supply	D 35	Working capital	1451
Construction	F	Buildings	19,044
Construction	F	Land preparation, enclosures, viability	1120
Construction	F	Construction expenses	11,491
Construction	F	Contingencies	7182
Real Estate activities	L	Land & Yard Improvements	213
Land transport	H 49	Purchased equipment delivery	2477

Table 6.12 Cost categories of TPC included in the upstream cutoff matrix for the operation phase (k€ of 2010)

ISIC Rev.4 reference sector—Code		Monetary cost category	Value (k€)
Manufacture of chemicals ...	C 20	Reagents for water treatment	48
Manufacture of chemicals ...	C 20	NaHCO3	778
Manufacture of chemicals ...	C 20	Activated carbon	9
Manufacture of chemicals ...	C 20	Lime	72
Manufacture of chemicals ...	C 20	NH3 solution	105
Manufacture of chemicals ...	C 20	Reagents for Fly-ash inertization	2
Extraction of crude petroleum and natural gas	B 06	Process methane	243
Manufacture of coke and refined products ...	C 19	Gasoil	10
Electric power generation, ...	D 35	Electricity	6
Water collection, treatment and supply	E 36	Process water	148
Repair of fabricated metal products, machinery	C 33	Maintenance and repair: equipment	990.7
Waste collection, treatment and disposal ...	E 38	Residue disposal	683.8
Construction of buildings	F 41	Maintenance and repair: building	571.3
Insurance, reinsurance and pension funding	K 65	Property Insurance	718.2

Results of the ELCA Analysis

Thanks to the economic analysis of the WtE plant, the Hybrid Input-Output system can be characterized for both construction and operation phases. Therefore, thanks to the application of the Leontief model, is it possible to account for the primary non-renewable exergy embodied in products of the WtE plant. This and the other indicators defined in Sect. 4.3.2 are resumed in Table 6.13. Results confirm the strong potential of Waste-to-Energy technology in the perspective of primary fossil fuel displacement. Based on the obtained results, the following considerations can be made. First of all, the exergy embodied in construction and operation phases equals the yearly total primary exergy supply of about nine hundred Italian citizens

Table 6.13 Results of ELCA analysis

Parameter	Parameter	Unit	Value
Primary exergy cost of Construction	$E_{H,Construction}$	toe	1745.9
Yearly primary exergy cost of Operation	$E_{H,Operation}$	toe/year	61.3
Yearly electricity production	Ex_P	toe/year	7062.5
Total primary exergy cost	E_H	toe	3278.3
Specific primary exergy cost	e_H	J/J	0.0186
Primary Net exergy cost	$E_{H,Net}$	ktoe	−173.3
Exergy return on investment	ExROI	–	99.25

in 2010, assuming about 2.8 toe pro-capita in 2010 (IEA 2010). Secondly, the WtE plant produces an overall amount of exergy (electricity) that exceeds the primary non-renewable resources required along its construction and operation phases: indeed, value of Net embodied exergy requirements is strongly negative, and specific primary exergy cost of electricity is less than one. Finally, value of the Exergy Return On Investment (ExROI) reveals that the analyzed system is able to produce a net amount of electricity about 100 times greater than the primary non-renewable resources required to own and operate it.

Disaggregated contributions of all the inputs taken from the Italian economy that are required for plant Construction and Operation can be calculated thanks to relation (4.41): detailed results for each phase are collected in Tables 6.14 and 6.15. For the construction phase, manufacture of machinery and equipment and construction of buildings represent more than 60 % of the total primary resources consumption. Surprisingly, more than 7 % of the primary exergy cost is devoted to non-material services (engineering and supervision), which are usually neglected in standard process-based LCA. Moreover, the largest fraction of the primary exergy cost of the operation phase (more than 50 %) is caused by raw materials supply and equipment maintenance. As above, non-material services play a non-negligible role, affecting primary exergy requirements for slightly less than 20 %.

6.2.4   Design evaluation and optimization of the system

Design evaluation and optimization procedure has been carried out based on the guidelines and the indicators introduced in Sect. 4.3.2. The objective of the optimization is the reduction of the exergy cost of the products of the WtE plant by

Table 6.14 Results of ELCA analysis: disaggregation of primary exergy cost of Construction

ISIC Rev.4 reference sector—Code		E_H [toe]	E_{NS} (%)
Manufacture of basic metals	C 24	121.1	6.9
Manufacture of electrical equipment	C 27	231.1	13.2
Manufacture of machinery and equipment n.e.c.	C 28	605.4	34.7
Repair and installation of machinery and equipment	C 33	127.1	7.3
Electricity, gas, steam, and air conditioning supply	D 35	17.7	1.0
Construction	F 41–43	474.6	27.2
Land transport and transport via pipelines	H 49	30.3	1.7
Real estate activities	L 68	2.6	0.1
Legal and accounting activities; activities of head offices; ...	M 69–70	8.8	0.5
Architectural and engineering activities; technical testing and analysis	M 71	127.1	7.3
Total for construction phase	$E_{H,Const.}$	**1745.9**	**100.0**

Table 6.15 Results of ELCA analysis: disaggregation of primary exergy cost of Operation

ISIC Rev.4 reference sector—Code		E_H (toe/year)	E_H (%)
Mining and quarrying	B 05–09	3.9	6.4
Manufacture of coke and refined petroleum product	C 19	0.2	0.3
Manufacture of chemicals and chemical products	C 20	16.4	26.7
Repair and installation of machinery and equipment	C 33	16.0	26.1
Electricity, gas, steam, and air conditioning supply	D 35	0.1	0.2
Water collection, treatment and supply	E 36	2.4	3.9
Sewerage; waste collection, treatment and disposal activities; ...	E 37–39	1.5	2.5
Construction	F 41–43	9.2	15.1
Insurance, reinsurance and pension funding, except compulsory ...	K 65	11.6	18.9
Total for operation phase	$E_{H,Oper.}$	**61.3**	**100.0**

focusing on components with the highest exergy cost of exergy destructions, acting on different key design parameters, namely: *grate furnace excess air ε* and *exhaust gas temperature T_8*, *steam turbine inlet temperature T_{28}* and *pressure p_{28}*, *air pre-heater steam pressure p_{29}*.

Any proposed design improvement based on traditional Exergy Cost Theory has been verified in a broader perspective using results of ELCA: to be effective, any change in the plant design should be associated to a reduction in the primary non-renewable exergy embodied in electricity production of the WtE plant. With the proposed ELCA, the primary exergy embodied in electricity production is based on the investment and operative costs of the system (TCI and TCP, collected in the Upstream cutoff matrix). Such values are defined as functions of the *Purchase Equipment Cost* (PEC) of the most relevant plant components, which in turn are derived as functions of the considered key design parameters. This practice allows to evaluate the primary exergy embodied in system products based on different design configurations of the plant. The PEC of WtE plant components have been calculated based on the following two methods:

- *Cost Functions.* Detailed cost functions can be retrieved in literature in the form of relation (6.6), allowing to account for the economic cost of the generic component *a* as a function of *k* design parameters β_k;

$$C_a = f(\beta_1, \ldots, \beta_k) \qquad (6.6)$$

- *Scaling.* The economic cost of the generic component results from the *six-tenths rule*, expressed by the function (6.7) (Peters et al. 2003): known cost value of a similar component *b* is scaled through the ratio between the most relevant design parameter β of the considered component. Usually, an exponential factor $\alpha < 1$ is adopted, depending on the type of the considered component. When more appropriate data are not available, $\alpha = 0.6$ is generally chosen;

Table 6.16 Results of the iterative design evaluation and optimization procedure

Sim.	Design parameter	Symbol	Unit	From	To	η_{ex} (%)	ExROI-
00	–					19.06	99.25
01	Grate furnace excess air	ε	%	40	30	19.13	99.70
02	Steam turbine inlet temperature	T_{28}	°C	390	420	19.64	102.20
03	Grate furnace exhaust gas temperature	T_8	°C	527	510	19.64	102.15
04	Steam turbine inlet pressure	p_{28}	bar	40	50	19.98	103.86
05	Underfire air pre-heater steam pressure	p_{29}	bar	12.6	11.0	20.0	103.91

$$C_a = C_b \cdot \left(\frac{\beta_a}{\beta_a}\right)^{\alpha} \tag{6.7}$$

With reference to Table 6.16, the starting design configuration (00) has been iteratively improved: changes in the key design parameters have been made based on the indications provided by the Exergy Cost Theory. For each new design configuration, results of ELCA independently quantify the overall benefits in terms of the overall primary exergy requirements of the WtE. The optimization process results in a reduction by about 4.5 % in both direct and primary exergy costs of electricity. In absolute values, this correspond to a direct exergy saving of about 487 toe (of Municipal Solid Waste) and a primary exergy saving of 7938 toe (of primary fossil fuels). Therefore, in this specific case study, reducing local irreversibilities results in an even greater decrease in primary exergy requirements.

6.3   Applications of the Bioeconomic ExIO analysis

In this section, the Bioeconomic ExIO model is applied the Italian economy in 2010, evaluating specific and total exergy embodied in national products. Results are compared with the ones obtained through the standard ExIO model (see Sect. 6.1.4). Finally, a comparative evaluation of alternative dishwashing practices in a Life Cycle perspective is carried out based on both standard and Bioeconomic ExIO models.

6.3.1   Primary exergy embodied in national economic production

The Bioeconomic and the standard ExIO models are here applied to the Italian economy in 2010. All the required data (MIOTs, primary exergy requirements, etc.) have been taken from the data sources listed in Sect. 6.1.

Bioeconomic ExIO model has been defined according to the guidelines presented in Sect. 5.2. For the sake of simplicity, international trades of products are treated according to Single-Region *model b* (see Sect. 4.1.2) for both standard and Bioeconomic ExIO models.

The Human Labour sector of the Italian economy has been characterized based on data retrieved in the WIOD database and in the related technical documentation (Erumban et al. 2012). Total population of Italy in 2010 was about 59.3 Millions, and the total amount of hours lived by the population h_{tot} were about 17.31×10^4 Mh. Working hours consumed by each sector of the economy are retrieved in the WIOD database (vector $\mathbf{h_W}$): based on the proportionality assumption defined by relation (5.6), fractions of working hours with respect to the total hours lived by the Italian population $(h_{W,i}/h_{tot})$, the total final demand $f_{tot,i}$, the total demand of households $f_{H,i}$ and the amount of final demand that sustains Human Labour production process $f_{H,W,i}$ are derived and listed in Table 6.17 for each ith Italian economic sector. Based on these values, the amount of final demand devoted to sustain workers $f_{H,W,i}$ results a non-negligible fraction of the total final demand $f_{tot,i}$. In few cases, the final demand devoted to sustain workers $f_{H,W,i}$ results greater than the household final demand $f_{H,i}$: this happens when sectors produces few final demand for household consuming high amounts of working hours (i.e. sectors C, 26, F, 71t74, L, M).

The environmental intervention vector $\mathbf{R_B}$ is defined as in relation (5.8) based on the *International Energy Agency* database. $\mathbf{R_B}$ includes only primary fossil fuels (raw coal, crude oil and natural gas) *endogenously produced* in year 2010 by the Italian economy, properly converted in *ktoe* of exergy. It is finally assumed that such amount of primary energy enters the Italian economy through the ISIC sector *Mining and quarrying* (code B).

Results of the analyses are collected in Table 6.18: specific ($\mathbf{e}$, in $kg_{oe}/100USD$) and total ($\mathbf{E}$, in ktoe) embodied exergy of each sector of the economy have been calculated according to both the Bioeconomic (subscript B) and the standard (subscript S) ExIO models. Relative differences between results (Δ) have been also calculated in percentage. Based on the obtained results, values of specific embodied exergy $\mathbf{e}$ calculated through the Bioeconomic model result always higher with respect to the results obtained by the standard model: this because the final demand of the national economy defined by the Bioeconomic model results as lower with respect to the standard model, due to the internalization of the Human Labour sector. However, the exergy embodied in products of the Italian economic sectors calculated according to the Bioeconomic and the standard model differs, but the total exergy embodied in Italian production (107,645 ktoe, last line of Table 6.18) is equal for both the models. This reveals that the total exergy embodied in national production must be equal to the total primary exergy requirements of the nation, and it can be established as a check for a correct application of the Bioeconomic model.

Two fundamental considerations can be made based on the obtained results: first, the application of the Bioeconomic model cause a reallocation of the exergy

Table 6.17 List of the considered economic sector (ISIC rev.3), with essential data for the application of the Bioeconomic Input-Output model. Year 2010

Code	$h_{W,i}$ (Mh)	$h_{W,i}/h_{tot}$ (%)	$f_{tot,i}$ (MUSD)	$f_{H,i}$ (MUSD)	$f_{H,W,i}$ (MUSD)
AtB	810	0.47	37,202	28,590	5420
C	57	0.03	2008	350	384
15t16	578	0.33	123,837	94,104	3868
17t18	688	0.40	77,877	45,825	4602
19	215	0.12	30,796	12,925	1437
20	174	0.10	6294	2701	1164
21t22	334	0.19	25,046	16,350	2235
23	40	0.02	46,322	25,755	267
24	321	0.19	84,387	24,550	2149
25	296	0.17	24,218	6905	1982
26	363	0.21	14,499	3939	2430
27t28	1275	0.74	82,312	6826	8531
29	987	0.57	133,806	11,637	6605
30t33	676	0.39	88,595	17,661	4520
34t35	423	0.24	95,678	29,201	2831
36t37	381	0.22	47,250	20,790	2551
E	191	0.11	42,089	40,331	1276
F	2173	1.26	183,833	10,982	14,539
50	628	0.36	57,540	48,585	4201
51	1092	0.63	91,969	57,810	7302
52	1664	0.96	88,741	61,925	11,130
H	1456	0.84	119,980	117,753	9741
60	663	0.38	53,411	40,466	4435
61	49	0.03	9165	4266	326
62	32	0.02	10,798	7052	216
63	644	0.37	26,513	15,496	4309
64	380	0.22	29,302	25,793	2544
J	890	0.51	67,090	57,829	5954
70	103	0.06	197,077	183,239	691
71t74	3344	1.93	69,676	18,964	22,371
L	2141	1.24	174,880	1533	14,319
M	2322	1.34	98,021	14,344	15,532
N	2339	1.35	168,185	29,404	15,645
O	1287	0.74	70,638	53,563	8608
P	2483	1.43	20,438	20,438	16,612
HL	0	0.00	0	0	0
Tot	**31,501**	**18**	**2,499,472**	**1,157,882**	**210,725**

Table 6.18 Specific (**e**) and total (**E**) exergy embodied in products of the Italian economy in 2010. Subscript "B" refers to the bioeconomic model, while subscript "S" refers to the standard model

Code	e_B (kg$_{oe}$/ 100USD)	e_S (kg$_{oe}$/ 100USD)	Δe (%)	E_B (ktoe)	E_S (ktoe)	ΔE (%)
AtB	3.44	3.13	10.2	1095	1163	−5.8
C	113.47	113.45	0.0	1842	2278	−19.1
15t16	3.43	3.19	7.4	4114	3954	4.0
17t18	3.00	2.72	10.1	2199	2122	3.6
19	2.16	1.90	13.5	635	586	8.2
20	3.00	2.69	11.5	154	169	−9.1
21t22	3.84	3.59	6.9	875	899	−2.6
23	64.33	64.27	0.1	29,628	29,770	−0.5
24	4.01	3.88	3.5	3302	3273	0.9
25	3.75	3.49	7.3	834	846	−1.5
26	14.77	14.48	2.0	1783	2100	−15.1
27t28	4.60	4.34	5.9	3396	3576	−5.0
29	2.86	2.59	10.4	3636	3463	5.0
30t33	2.01	1.81	10.9	1692	1607	5.3
34t35	2.20	2.01	9.6	2042	1920	6.3
36t37	2.75	2.47	11.5	1229	1166	5.4
E	38.88	38.78	0.3	15,867	16,321	−2.8
F	4.30	3.97	8.5	7286	7291	−0.1
50	2.66	2.36	12.7	1419	1359	4.5
51	2.69	2.46	9.5	2280	2263	0.8
52	3.79	3.46	9.5	2940	3070	−4.2
H	3.04	2.72	11.7	3346	3259	2.7
60	4.98	4.78	4.2	2440	2553	−4.4
61	3.80	3.51	8.2	336	322	4.3
62	6.21	6.05	2.7	657	653	0.6
63	2.98	2.64	13.1	662	700	−5.3
64	2.10	1.87	12.2	561	548	2.4
J	1.01	0.81	23.9	617	546	12.9
70	0.44	0.41	6.8	861	808	6.5
71t74	2.06	1.74	18.4	976	1214	−19.6
L	2.05	1.71	20.0	3286	2982	10.2
M	1.28	0.76	68.3	1055	745	41.6
N	1.97	1.60	23.7	3010	2683	12.2
O	2.40	2.03	17.9	1487	1436	3.6
P	2.74	0.00	100.0	105	0	100.0
HL	22.55	nd	100.0	0	nd	100.0
Tot	–	–	–	**107,645**	**107,645**	–

embodied in goods and services that takes into account their labour embodiment. Secondly, the evaluation of embodied exergy strongly depends on the definition of the final demand, that is, on what is considered as the *useful output* of any economic process.

Changes in values of embodied exergy of products caused by the internalization of the Human Labour sector within the national economy is ultimately due to two factors: (1) the change in technology and (2) the reduction of the national final demand. The effects of these two structural changes in the national economy can be decoupled and quantified by means of the *Structural Decomposition Analysis* (SDA) (Nakamura and Kondo 2009), described in Sect. 2.3.6 and here applied with reference to relation (6.8).

$$\Delta E = \Delta E_t + \Delta E_f = \underbrace{\hat{f}_B \cdot (e_B - e_S)}_{technology} + \underbrace{\left(\hat{f}_B - \hat{f}_S\right) \cdot e_S}_{final\ demand} \qquad (6.8)$$

Table 6.19 resumes the results of the SDA: any change in technology results always in positive changes of embodied exergy in products, proportional to the amount of working hours absorbed by each sector of the economy. Conversely, the effect of the final demand reduction (ΔE_f) are always negative and proportional to the amount of final demand internalized in the economy by the Human Labour sector. The algebraic sum of these two effects return the net change in embodied exergy of each sector (ΔE_{tot}). Because of these two overlapped effects, it is not always said that sectors that invoke greater amount of working hours result in higher values of embodied exergy. These overlapped effects make results of the Bioeconomic model non trivial, since there is not a direct relation between working hours consumption of each sector and the change in embodied exergy of its production.

Finally, notice that the Human Labour production process is characterized by a value of embodied exergy of 22.25 kg$_{oe}$/100 h (see Table 6.18). This value can be used as a reference for the embodied exergy in the average working hour produced by the Italian economy. However, since working hours are not part of the final demand, total embodied exergy of the working hours production sector turns out to be zero. Therefore, the Bioeconomic model cause a reallocation of primary exergy requirements among all the productive sectors of the economy.

6.3.2 Primary exergy requirements of alternative dishwashing practices

In the evaluation of environmental burdens of economic activities, the primary exergy requirements of household appliances in a life cycle perspective is under close scrutiny (Aydinalp et al. 2002; Geller et al. 2006; Ueno et al. 2006). Here, both the standard and the Bioeconomic ExIO models are applied for the analysis of *manual dishwashing* (HW) and *dishwasher* (DW) practices. Indeed, dishwashers

Table 6.19 Results of the SDA. "ΔE_t" refers to the effects due to changes in technology, while "ΔE_f" refers to the effects due to changes in final demand

Code	ΔE_t (ktoe)	ΔE_f (ktoe)	ΔE_{tot} (ktoe)
AtB	101	−169	−68
C	0	−436	−435
15t16	283	−124	159
17t18	202	−125	77
19	76	−27	48
20	16	−31	−15
21t22	57	−80	−23
23	29	−172	−142
24	112	−83	29
25	57	−69	−12
26	35	−352	−317
27t28	191	−371	−180
29	344	−171	173
30t33	167	−82	85
34t35	179	−57	122
36t37	126	−63	63
E	41	−495	−454
F	572	−577	−5
50	160	−99	61
51	197	−180	17
52	255	−385	−130
H	352	−265	87
60	99	−212	−113
61	25	−11	14
62	17	−13	4
63	76	−114	−37
64	61	−48	13
J	119	−48	70
70	55	−3	52
71t74	152	−390	−238
L	548	−244	304
M	428	−118	310
N	576	−250	327
O	226	−175	51
P	105	0	105
HL	0	0	0

are used to substitute or supplement manual dishwashing, so the question arises not only for the environmental effects of automatic dishwashers but also of manual dishwashing. Indeed, while on the water consumption there is a clear advantage of

Table 6.20 Inventory analysis of the alternative dishwashing practice in Italy. Average values

Data	Unit	HW	DW
Washer	USD/unit	0	1000
Water	l/wash	150	15
Electric energy	kWh/wash	–	1.5
Natural gas	kWh/wash	3.5	–
Detergents	g/wash	50	30
Human labour	min/wash	30	10

using a dishwasher, evaluation of primary exergy requirements is more complex: exergy embodied in detergents, water, electricity, natural gas, etc. should to be properly taken into account (Berkholz et al. 2010), as well as the different amount of human labour invoked by the two alternatives.

The analysis has been carried out in the Italian context, assuming the average washing requirements of a family for a total number of 300 days per year, and a total time window of 10 years. Average prices for dishwashers in 2010 (1000 USD/unit), water (0.1 cUSD/l), electricity (15 cUSD/kWh), natural gas (8 cUSD/kWh) and detergents (7 USD/kg) have been assumed. Human labour requirements (in terms of working hours), and water, energy and detergents monetary expenditures for the two alternatives have been collected in Table 6.20. All these data have been retrieved from dedicated literature (Berkholz et al. 2010; Brückner and Stamminger 2015; Stamminger et al. 2003), European regulations (European Committee for Electrotechnical Standardization 2003) and also from Italian statistics reports. The Italian MIOT of 2010 has been adopted as reference for the analysis based on the WIOD database.

Results of both standard and Bioeconomic analyses are presented in Table 6.21 and in Fig. 6.7. In both cases, manual dishwashing (HW) results as the most

Table 6.21 Detailed numerical results of standard and bioeconomic IO models in kg_{oe}

Input	Sector	Code	$E_{B,HW}$ (kg_{oe})	$E_{B,DW}$ (kg_{oe})	$E_{S,HW}$ (kg_{oe})	$E_{S,DW}$ (kg_{oe})
Washer	Machinery, nec	29	–	29	–	26
Water	Electricity, gas and water supply	E	175	17	174	17
Electricity	Electricity, gas and water supply	E	–	262	–	262
Natural gas	Electricity, gas and water supply	E	335	–	334	–
Detergents	Chemicals and chemical products	24	42	25	41	24
Labour	Human labour	HL	338	113	–	–
Total			**890**	**447**	**549**	**330**

S standard model, *B* bioeconomic model, *HW* manual dishwashing, *DW* dishwasher

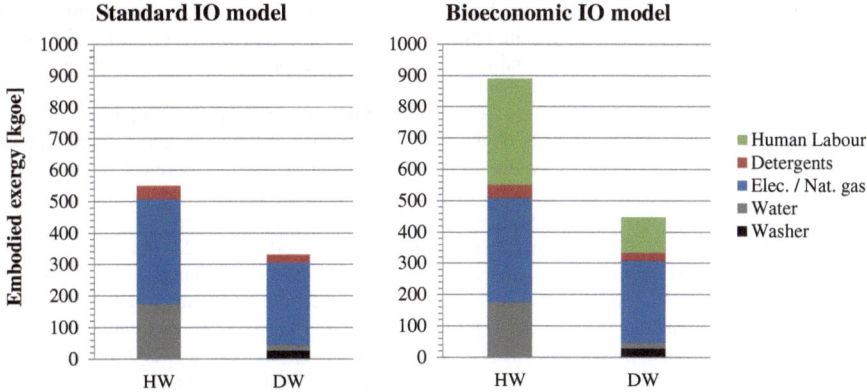

Fig. 6.7 Embodied exergy of manual dishwashing (HW) and dishwasher (DW) evaluated through the standard (*left side*) and Bioeconomic (*right side*) Input-Output models

primary exergy intensive practice. Moreover, the difference in primary exergy requirements between HW and DW is emphasized using the Bioeconomic model. Primary exergy requirements of the single material/energy inputs in HW and DW is not significantly affected switching from standard to Bioeconomic model. On the other hand, contributions of human labour is relevant for manual dishwashing (HW) only, causing almost a doubling of the primary exergy requirements with respect to the standard model. As expected, dishwasher (DW) does not show the same trend, since less human labour is involved. With both standard and Bioeconomic approaches, the exergy embodied in the dishwasher and in detergents represent small contributions with respect to the total exergy requirements.

References

Aydinalp, M., Ugursal, V. I., & Fung, A. S. (2002). Modeling of the appliance, lighting, and space-cooling energy consumptions in the residential sector using neural networks. *Applied Energy, 71,* 87–110.

Berkholz, P., Stamminger, R., Wnuk, G., Owens, J., & Bernarde, S. (2010). Manual dishwashing habits: An empirical analysis of UK consumers. *International Journal of Consumer Studies, 34,* 235–242.

Bohm, R. A., Folz, D. H., Kinnaman, T. C., & Podolsky, M. J. (2010). The costs of municipal waste and recycling programs. *Resources, Conservation and Recycling, 54,* 864–871.

Brückner, A., & Stamminger, R. (2015). Consumer-relevant assessment of automatic dishwashing machines by a new testing procedure for 'automatic' programmes. *Energy Efficiency, 8,* 171–182.

Buonanno, G., Ficco, G., & Stabile, L. (2009). Size distribution and number concentration of particles at the stack of a municipal waste incinerator. *Waste Management, 29,* 749–755.

Consonni, S. (2012). *Energy recovery 8: Economics of waste-to-energy, sustainable waste management.* AIT Athens: MatER.

Coronado, C. R., Tuna, C. E., Zanzi, R., Vane, L. F., & Silveira, J. L. (2013). Development of a thermoeconomic methodology for the optimization of biodiesel production—Part I: Biodiesel plant and thermoeconomic functional diagram. *Renewable and Sustainable Energy Reviews, 23*, 138–146.

Dietzenbacher, E., & Lahr, M.L. (2004). Wassily Leontief and input-output economics.

Dietzenbacher, E., Los, B., Stehrer, R., Timmer, M., & De Vries, G. (2013). The construction of world input–output tables in the WIOD project. *Economic Systems Research, 25*, 71–98.

Erumban, A., Gouma, R., de Vries, G., de Vries, K., & Timmer, M. (2012). WIOD socio-economic accounts (SEA): Sources and methods. *Groningen, April.*

European Committee for Electrotechnical Standardization, B., (2003). N 50242 A1:2003. Electric Dishwashers for Household Use—Methods for Measuring the Performance.

Eurostat (2008). NACE rev.2. Statistical classification of economic activities in the European Community. *Luxembourg: Office for Official Publications of the European Communities.*

Geller, H., Harrington, P., Rosenfeld, A. H., Tanishima, S., & Unander, F. (2006). Polices for increasing energy efficiency: Thirty years of experience in OECD countries. *Energy Policy, 34*, 556–573.

Genty, A., Arto, I., & Neuwahl, F. (2012). Final database of environmental satellite accounts: Technical report on their compilation. *WIOD Documentation.*

Glucina, M. D., & Mayumi, K. (2010). Connecting thermodynamics and economics. *Annals of the New York Academy of Sciences, 1185*, 11–29.

Green, D., & Perry, R. H. (1999). Perry's Chemical Engineers' Handbook. 7 edition.

Heijungs, R., & Suh, S. (2002). The computational structure of life cycle assessment. Springer.

Hendrickson, C. T., Horvath, A., Joshi, S., Klausner, M., Lave, L. B., McMichael, F. C. (1997). Comparing two life cycle assessment approaches: a process model vs. economic input-output-based assessment. In *Proceedings of the 1997 IEEE International Symposium on Electronics and the Environment, 1997.* ISEE-1997 (pp. 176–181). IEEE.

Hendrickson, C. T., Lave, L. B., & Matthews, H. S. (2010). Environmental life cycle assessment of goods and services: An input-output approach.

IEA. (2010). Italy: balances for 2010. *Key World Energy Statistics.*

IEA. (2014). Key World Energy Statistics 2014.

Kostowski, W. J., Usón, S., Stanek, W., & Bargiel, P. (2014). Thermoecological cost of electricity production in the natural gas pressure reduction process. *Energy*, 1–9.

Lave, L., MacLean, H., Hendrickson, C., & Lankey, R. (2000). Life-cycle analysis of alternative automobile fuel/propulsion technologies. *Environmental Science and Technology, 34*, 3598–3605.

Massarutto, A., Carli, A. D., & Graffi, M. (2011). Material and energy recovery in integrated waste management systems: A life-cycle costing approach. *Waste Management 31*, 2102–2111.

Mayumi, K. (2009). Nicholas Georgescu-Roegen: His bioeconomics approach to development and change. *Development and Change, 40*, 1235–1254.

Nakamura, S., & Kondo, Y. (2009). Waste input-output analysis: concepts and application to industrial ecology. Springer.

Peters, G. P. (2007). Efficient algorithms for life cycle assessment, input-output analysis, and Monte-Carlo analysis. *International Journal of Life Cycle Assessment, 12*, 373–380.

Peters, M. S., Timmerhaus, K. D., West, R. E., Timmerhaus, K., & West, R. (2003). Plant design and economics for chemical engineers.

Song, G., Xiao, J., Zhao, H., & Shen, L. (2012). A unified correlation for estimating specific chemical exergy of solid and liquid fuels. *Energy, 40*, 164–173.

Stamminger, R., Badura, R., Broil, G., Dorr, S., Elschenbroich, A., & Dörr, S. et al. (2003). A European Comparison of cleaning dishes by hand. In *Proceedings of EEDAL Conference* (pp. 735–743).

Szargut, J., Valero, A., Stanek, W., & Valero, A. (2005a). Towards an international legal reference environment. *Proceedings of ECOS, 2005*, 409–420.

Szargut, J., Valero, A., Stanek, W., & Valero, A. (2005b). *Towards an international reference environment of chemical exergy.* Preprint: Elsevier Science.

Tecnoborgo (2013). Dichiarazione ambientale, *Piacenza*.

Timmer, M. P., Dietzenbacher, E., Los, B., Stehrer, R., & de Vries, G. J. (2015). An illustrated user guide to the world input-output database: The case of global automotive production. *Review of International Economics*, n/a-n/a.

Timmer, M., Erumban, A. A., Gouma, R., Los, B., Temurshoev, U., de Vries, G. J., Arto, I. (2012). The world input-output database (WIOD): contents, sources and methods. *WIOD Background document available at* www.wiod.org 40.

Torres Cuadra, C., Valero, A., Rangel, V., & Zaleta, A. (2008). On the cost formation process of the residues. *Energy, 33*, 144–152.

Torres, C., Valero, A., & Valero, A. (2013). Exergoecology as a tool for ecological modelling. The case of the US food production chain. *Ecological Modelling, 255*, 21–28.

Tsilemou, K., & Panagiotakopoulos, D. (2006). Approximate cost functions for solid waste treatment facilities. *Waste Management and Research, 24*, 310–322.

Tukker, A., Huppes, G., Oers, L. V., & Heijungs, R. (2006). Environmentally extended input-output tables and models for Europe.

Ueno, T., Sano, F., Saeki, O., & Tsuji, K. (2006). Effectiveness of an energy-consumption information system on energy savings in residential houses based on monitored data. *Applied Energy, 83*, 166–183.

United Nations, European Commission, International Monetary Fund, Organisation for Economic Co-operation and Development, World Bank (2009). System of National Accounts 2008. UN.

Usón, S., & Valero, A. (2011). Thermoeconomic diagnosis for improving the operation of energy intensive systems: Comparison of methods. *Applied Energy, 88*, 699–711.

Valero, A., Usón, S., Torres, C., Valero, A., Agudelo, A., & Costa, J. (2013). Thermoeconomic tools for the analysis of eco-industrial parks. *Energy, 62*, 62–72.

Watkins, D. S. (2004). Fundamentals of matrix computations. Wiley.

Chapter 7
Conclusions

In this final chapter, advantages and drawbacks of the ExIO framework are highlighted, and possible further research paths are proposed.

7.1 Advantages and drawbacks of the ExIO framework

The ExIO framework has been developed to deal with the main issues emerging from the literature, highlighted in the introduction and assumed as the main objectives of this book. The most relevant achievements of the book are resumed below, together with the main advantages and drawbacks of the ExIO framework.

Identification of a standardized energy-resources accounting method. As extensively showed in Chap. 4, mathematical formulation of the Exergy-based Input-Output framework is based on the Input-Output analysis, establishing standard rules for the definition of time and space boundaries of any analyzed productive system, encompassing its whole Life Cycle. The approach relies on *Monetary Input Output Tables* of national economies, a freely available and constantly updated data source, which comprehensively classifies all the production activities within the economy according to international standards. As confirmed by the case studies presented in Chap. 6, the ExIO framework avoids extensive data mining processes, allowing to perform *reproducible* and *reliable* evaluations of primary energy-resources embodied in products. Moreover, application of such technique results to be *simpler* and *faster* with respect to other traditional process-based life cycle methods. Beside these advantages, the use of MIOTs requires suited hypotheses and models for the treatment of international trades of products. The most common models available in the literature have been comprehensively formalized in Chap. 4. As showed by the practical applications of ExIO to the case study of Sect. 6.1, an appropriate choice of the international trades model is critical and could largely affect the accuracy of results.

M.V. Rocco, *Primary Exergy Cost of Goods and Services*,
PoliMI SpringerBriefs, DOI 10.1007/978-3-319-43656-2_7

One other critical issue of the ExIO analysis concerns the low accuracy of results. Indeed, only few multi-directional Input-Output models are nowadays available, and they are compiled with coarse aggregation of economic activities. Therefore, MIOTs are suited to understand and to analyze the trades in embodied exergy among national economies, but they can be hardly adopted to perform LCA of detailed products. The *Hybrid-ExIO model* has been proposed and formalized to selectively extract the analyzed system from the economic sector in which it operates. This method results very useful in order to perform LCA analysis of detailed systems, as showed by the analyses of the Waste-to-Energy plant and dish cleaning options performed in Chap. 6. However, the application of such method requires detailed economic costs data related to the analyzed system.

One of the most relevant advantages of ExIO framework resides in the opportunity to perform both *Exergy Cost Analysis* and *Exergy Life Cycle Assessment* of energy conversion systems, using results of both methods to perform design evaluation and optimization of energy conversion systems. The Hybrid-ExIO turns out to be a suited approach to perform ELCA, allowing to define quantitative criteria and suited indicators to identify and to minimize the primary fossil fuels requirements of the products of energy conversion systems in a Life Cycle perspective.

From the methodological standpoint, the Hybrid-ExIO model reveals to be a simple, standardized and promising technique to perform LCA of energy systems: however, more applications are required to test and to improve such method.

Because of its features, application of the ExIO analysis is affected by uncertainties propagation that could negatively influence the quality and the accuracy of results. Due to the mathematical structure of ExIO, uncertainty analysis could be efficiently performed by means of *Perturbation Theory*: this issue has not been addressed in this book, and it may be considered as one of the possible further research paths.

Characterization of non-renewable energy-resources based on thermodynamics. With respect to the other thermodynamic-based metrics, exergy represents the real usefulness of energy and bulk flow interactions, resulting as the most meaningful indicator for energy-resources accounting purpose. Moreover, exergy of fossil fuels seems to be less sensible to the conditions of reference environment with respect to the entropy generation. For such reasons, exergy is assumed by the ExIO framework as the unique unit to account for the primary energy-resources requirements of goods and services.

Role of working hours. The *Bioeconomic ExIO model* has been proposed as a partially closed Input-Output model able to internalize the effects of human labour into the primary energy-resources accounting. In general, the partial internalization of the final demand performed by the Bioeconomic model cause a reallocation of the primary energy-resources requirements among all the productive sectors of the economy: this allows to take into account for the additional goods and services consumption required to feed workers that produce working hours, causing an increase of the total exergy embodied in products out of tertiary sector with respect to other less working hours intensive sectors. Both the Standard and the

Bioeconomic ExIO models have been used to compare manual dishwashing and dishwasher practices within the Italian context, evaluating the total primary exergy requirements of both solutions in a LCA perspective. From the methodological standpoint, the amount of final demand products that are actually feeding the working hours production process is determined according to a simple proportionality hypothesis, introduced in Chap. 5. Further research is required in order to test this hypothesis or to propose a more refined criteria.

7.2   Final Remarks

The ExIO framework allows to quantify the primary resources directly and indirectly devoted to the production of one defined product. Therefore, ExIO enables to understand the purpose for which primary resources are actually depleted, providing useful information to define efficient policies and technical improvements of supply chains.

As can be inferred from the literature and from the results obtained in this book, production of goods and services in modern economies cause relevant *indirect* energy-resources requirements that are ignored by conventional thermodynamics-based analyses. Without a proper evaluation of the *overall* resource consumption of one specific productive system, capable to include also the indirect supply chains requirements and externalities effects, misleading results may be obtained and wrong political and technological decisions may be taken.

The *Exergy-based Input-Output* framework proposed in this book aims to provide analysts and policymakers a comprehensive and useful tool for the evaluation of primary energy-resources requirements of goods and services, and for the analysis and the optimization of energy conversion systems in a Life Cycle perspective.